Springer Tracts in Advanced Robotics

Volume 61

Óscar Martínez Mozos

Semantic Labeling of Places with Mobile Robots

Springer

Professor Bruno Siciliano, Dipartimento di Informatica e Sistemistica, Università di Napoli Federico II, Via Claudio 21, 80125 Napoli, Italy, E-mail: siciliano@unina.it

Professor Oussama Khatib, Artificial Intelligence Laboratory, Department of Computer Science, Stanford University, Stanford, CA 94305-9010, USA, E-mail: khatib@cs.stanford.edu

Professor Frans Groen, Department of Computer Science, Universiteit van Amsterdam, Kruislaan 403, 1098 SJ Amsterdam, The Netherlands, E-mail: groen@science.uva.nl

Author

Dr. Óscar Martínez Mozos
Autonomous Intelligent Systems
Department of Computer Science
University of Freiburg
Georges-Koehler-Allee, Geb. 079
79110 Freiburg
Germany
E-mail: omozos@gmail.com

ISBN 978-3-642-11209-6 e-ISBN 978-3-642-11210-2

DOI 10.1007/978-3-642-11210-2

Springer Tracts in Advanced Robotics ISSN 1610-7438

Library of Congress Control Number: 2009942892

Typeset & Cover Design: Scientific Publishing Services Pvt. Ltd., Chennai, India.

Printed in acid-free paper

5 4 3 2 1 0

springer.com

STAR (Springer Tracts in Advanced Robotics) has been promoted under the auspices of EURON (European Robotics Research Network)

To my family and friends
A mi familia y amigos

Editorial Foreword

By the dawn of the new millennium, robotics has undergone a major transformation in scope and dimensions. This expansion has been brought about by the maturity of the field and the advances in its related technologies. From a largely dominant industrial focus, robotics has been rapidly expanding into the challenges of the human world. The new generation of robots is expected to safely and dependably co-habitat with humans in homes, workplaces, and communities, providing support in services, entertainment, education, healthcare, manufacturing, and assistance.

Beyond its impact on physical robots, the body of knowledge robotics has produced is revealing a much wider range of applications reaching across diverse research areas and scientific disciplines, such as: biomechanics, haptics, neurosciences, virtual simulation, animation, surgery, and sensor networks among others. In return, the challenges of the new emerging areas are proving an abundant source of stimulation and insights for the field of robotics. It is indeed at the intersection of disciplines that the most striking advances happen.

The goal of the series of *Springer Tracts in Advanced Robotics* (STAR) is to bring, in a timely fashion, the latest advances and developments in robotics on the basis of their significance and quality. It is our hope that the wider dissemination of research developments will stimulate more exchanges and collaborations among the research community and contribute to further advancement of this rapidly growing field.

The monograph written by Óscar Martínez Mozos is a contribution in the area of mobile robot navigation and perception, which has been receiving a great deal of attention by the research community in the latest few years. The contents expand the author's doctoral dissertation and are focused on the problem of assigning semantic labels to sensor data. A number of solutions inherited from pattern recognition and classification are introduced to enable a mobile robot to categorize places in typical indoor environments, such as a room or a corridor. Results are accompanied by a rich set of experiments, revealing a promising outlook toward the application to a wide range of mobile robots and service settings, such as elderly care, guiding, office and domestic assistance, and inspection.

Yet another STAR volume on mobile robots, a very fine addition to the series!

Naples, Italy
October 2009

Bruno Siciliano
STAR Editor

Foreword

Service robots that are able to operate autonomously and to interact with their users are one of the major research goals in robotics. In previous years there has been a tremendous progress in the area of robot navigation and perception. However, the question of how to bridge the gap between such subsymbolic and symbolic tasks like planning or interaction has remained an open research question.

In this book, Óscar Martínez Mozos addresses the problem of assigning semantic labels to sensor data. He considers problems like labeling places in the environment, extracting topological representations from geometric maps, utilizing semantic labels for exploration and navigation as well as for human-robot interaction. All approaches have been validated on real robotic systems and on data gathered with real robots.

To my opinion, this book describes an approach that builds the basis of a roadmap towards getting from subsymbolic sensations to symbolic representations. It describes methods for the direct interpretation and also for the temporal and spatial integration. I truly believe that it will be the basis for plenty of follow-up research projects that build upon the approaches described in this work and take us closer to semantically embedded robots.

Freiburg, Germany
September 2009

Wolfram Burgard

Preface

During the last years there has been an increasing interest in the area of service robots. Under this category we find robots working in tasks such as elderly care, guiding, office and domestic assistance, inspection, and many more.

Service robots usually work in indoor environments designed for humans, with offices and houses being some of the most typical examples. These environments are typically divided into places with different functionalities like corridors, rooms or doorways. The ability to learn such semantic categories from sensor data enables a mobile robot to extend its representation of the environment, and to improve its capabilities. As an example, natural language terms like *corridor* or *room* can be used to indicate the position of the robot in a more intuitive way when communicating with humans.

This book presents several approaches to enable a mobile robot to categorize places in indoor environments. The categories are indicated by terms which represent the different regions in these environments. The objective of this work is to enable mobile robots to perceive the spatial divisions in indoor environments in a similar way as people do. This is an interesting step forward to the problem of moving the perception of robots closer to the perception of humans.

Many approaches introduced in this book come from the area of pattern recognition and classification. The applied methods have been adapted to solve the specific problem of place recognition. In this regard, this work is a useful reference to students and researchers who want to introduce classification techniques to help solve similar problems in mobile robotics. In addition, the data sets corresponding to most of the experiments in the book can be found at

http://www.informatik.uni-freiburg.de/~omartine/place_data_sets.html

This book was possible thanks to the help and support of many people. First of all, I would like to thank Wolfram Burgard for giving me the opportunity to work under his supervision. Many thanks to Aleš Leonardis for his interest in my work. And thanks to Henrik Christensen for supporting this book. Special thanks to friends and colleagues that contributed to this book: Cyrill Stachniss, Axel Rottmann, Rudolph Triebel, Kai Arras, Hendrik Zender, and Patric Jensfelt.

I would like to thank the people providing the data sets which I used in my experiments. Cyrill Stachniss provided the map of building 79 at the University of Freiburg. Steffen Gutmann provided the map of building 52 at the University of Freiburg. Dieter Fox provided the map of the interior of the Intel Research Lab in Seattle. Finally, Andrew Howard provided the map of the building in Virginia.

This work is the result of my stay as a PhD student in the Autonomous Intelligent Systems Laboratory in Freiburg. I would like to thank the staff of this laboratory for their support.

During my stay in Freiburg I met many interesting people and made very good friends. I would like to thank all of them for making my life in this city more comfortable.

In addition, I would like to thank Justin Bachmann for helping me with the proofreading of the book.

Mamá, papá, hermano y resto de familia, vosotros sois los más importantes.

Alicante, Spain
September 2009

Óscar Martínez Mozos

Contents

1 **Introduction** ... 1
References ... 5

2 **Supervised Learning** ... 7
2.1 Introduction ... 7
2.2 Boosting ... 9
2.2.1 AdaBoost ... 10
2.2.2 Generalized AdaBoost ... 12
References ... 13

3 **Semantic Learning of Places from Range Data** ... 15
3.1 Introduction ... 15
3.2 Binary Classification Using AdaBoost ... 17
3.3 Classification of Multiple Classes ... 18
3.4 Simple Features from Sensor Range Data ... 19
3.5 Feature Extraction with Restricted Field of View ... 22
3.6 Experimental Results ... 22
3.6.1 Results Using Decision Lists ... 23
3.6.2 Transferring the Classifiers to New Environments ... 27
3.6.3 Place Recognition with a Moving Robot ... 29
3.6.4 Classification of Trajectories Using Sensors with Restricted Field of View ... 30
3.6.5 Selected Features ... 31
3.7 Related Work ... 32
3.8 Conclusion ... 33
References ... 33

4 **Topological Map Extraction with Semantic Information** ... 35
4.1 Introduction ... 35

4.2 Generalized AdaBoost 37
4.3 Probabilistic Decision List 39
4.4 New Geometrical Features from Sensor Range Data 40
4.5 Probabilistic Relaxation Labeling 41
4.6 Instance-Based Associative Markov Networks 43
4.6.1 Associative Markov Networks 43
4.6.2 Feature Vector Transformation 44
4.6.3 Feature Selection 46
4.7 Region Extraction and Topological Mapping 47
4.8 Experimental Results 48
4.8.1 Results Using Relaxation Labeling 48
4.8.2 Application to New Indoor Environments 51
4.8.3 Results with Instance-Based Associative Markov Networks 51
4.8.4 Comparison of Feature Sets 53
4.9 Related Work 54
4.10 Conclusion 55
References 55

5 Probabilistic Semantic Classification of Trajectories 57
5.1 Introduction 57
5.2 Generalized AdaBoost 58
5.3 Simple Features from Laser and Vision Data 59
5.4 Probabilistic Trajectory Classification 59
5.5 Experimental Results 62
5.5.1 Classifying Places along Trajectories 63
5.5.2 Improvement Obtained by Combining Laser and Vision Data 67
5.6 Related Work 68
5.7 Conclusion 68
References 69

6 Semantic Information in Exploration and Localization 71
6.1 Introduction 71
6.2 Multi-robot Exploration Using Semantic Information 72
6.2.1 Classifying Target Locations 72
6.2.2 Target Assignment Using Semantic Place Labeling 73
6.3 Localization Using Place Recognition 74
6.4 Experimental Results 75
6.4.1 Multi-robot Exploration 75
6.4.2 Localization 78
6.5 Related Work 78
6.6 Conclusion 79
References 80

7 Conceptual Spatial Representation of Indoor Environments 83
7.1 Introduction 83
7.2 Multi-layered Conceptual Mapping 84
7.2.1 Metric Map 85
7.2.2 Navigation Map 85
7.2.3 Topological Map 87
7.2.4 Conceptual Map 87
7.3 System Integration 89
7.4 Demo 92
7.5 Related Work 94
7.6 Conclusion 95
References 95

8 Semantic Information in Sensor Data 99
8.1 Introduction 99
8.2 Feature Extraction 101
8.3 Classification Using Boosting 102
8.4 Experimental Results 103
8.4.1 Corridor and Office Environments 103
8.4.2 Transferring the Classifiers to New Environments 105
8.4.3 Experiments Including the Motion Feature 105
8.4.4 Best Features for People Detection 106
8.5 Related Work 106
8.6 Conclusion 107
References 108

9 Conclusion 109

A Simple Features for Place Classification Using Raw Laser Beams ... 113
A.1 The Average Difference between the Length of Two Consecutive Beams 113
A.2 The Standard Deviation of the Difference between the Length of Two Consecutive Beams 114
A.3 The Average Difference between the Length of Two Consecutive Beams Considering Max-Range Values 114
A.4 The Standard Deviation of the Difference between the Length of Two Consecutive Beams Considering Max-Range Values 114
A.5 The Average Beam Length 114
A.6 The Standard Deviation of the Beam Length 115
A.7 Number of Gaps 115
A.8 The Number of Beams Lying on Lines Extracted from the Range Scan 116
A.9 The Euclidean Distance between the Two Points Corresponding to Two Global Minima 116

A.10 The Angular Distance between the Two Points Corresponding to Two Global Minima ... 116
A.11 The Average of the Relation between the Lengths of Two Consecutive Beams ... 116
A.12 The Standard Deviation of the Relation between the Length of Two Consecutive Beams ... 117
A.13 The Average of the Normalized Beam Length ... 118
A.14 The Standard Deviation of the Normalized Beam Length ... 118
A.15 The Number of Relative Gaps ... 118
A.16 Kurtosis ... 118
References ... 119

B Simple Features for Place Classification ... 121
B.1 Area of Pol(z) ... 121
B.2 Perimeter of Pol(z) ... 122
B.3 The Area of Pol(z) Divided by Its Perimeter ... 122
B.4 The Mean Distance between the Centroid and the Shape Boundary of Pol(z) ... 122
B.5 The Standard Deviation of the Distances between the Centroid and the Shape Boundary of Pol(z) ... 122
B.6 Similarity Invariant Descriptors Based on the Fourier Transformation of Pol(z) ... 123
B.7 Major Axis of the Ellipse That Approximates Pol(z) ... 123
B.8 Minor Axis of the Ellipse That Approximates Pol(z) ... 124
B.9 Invariant Moments of Pol(z) ... 124
B.10 The Normalized Feature of Compactness of Pol(z) ... 125
B.11 The Normalized Feature of Eccentricity of Pol(z) ... 126
B.12 The Form Factor of Pol(z) ... 126
B.13 Circularity of Pol(z) ... 126
B.14 The Normalized Circularity of Pol(z) ... 126
B.15 The Average Normalized Distance between the Centroid and the Shape Boundary of Pol(z) ... 127
B.16 The Standard Deviation of the Normalized Distances between the Centroid and the Shape Boundary of Pol(z) ... 127
References ... 127

C Simple Features for People Detection ... 129
C.1 Number of Points in the Segment ... 129
C.2 The Standard Deviation of the Beam Length ... 129
C.3 The Mean Average Deviation from the Median ... 129
C.4 Jump Distance from Preceding Segment ... 130
C.5 Jump Distance to Succeeding Segment ... 130
C.6 Euclidian Distance between the First and Last Point of a Segment ... 130
C.7 The Linearity of the Segment ... 130
C.8 The Circularity of the Segment ... 131

C.9 The Radius of the Circle Fitted to the Segment 131
C.10 The Boundary Length of the Segment 131
C.11 The Boundary Regularity of the Segment....................... 131
C.12 The Average Curvature of the Segment 132
C.13 The Mean Angular Difference of the Segment.................. 132
C.14 Mean Speed between Two Consecutive Scans 132
References 133

Notation

α	weak classifier weight
ε	weak classifier error
τ	feature transformation
$bel(\cdot)$	belief
f	feature
$h(\cdot)$	weak classifier
$p(k)$	probability of class k
x	example
y_k	label of class k
z	robot observation
A	set of features from the raw beams of an observation z
B	set of features from a polygonal approximation $\mathrm{Pol}(z)$ of an observation z
C^+	confidence value for a positive classification
C^-	confidence value for a negative classification
D	weight distribution over the training examples
E	set of edges in a graph, $\mathrm{E} \subseteq \mathrm{V} \times \mathrm{V}$
G	undirected graph
$H(\cdot)$	strong classifier
K	number of elements in Y
L	total number of features representing an example x
N	number of training examples
$\mathrm{Ne}(\cdot)$	neighborhood of a node in a graph
P	probability distribution over places
O	set of influence weights of nodes
$\mathrm{Pol}(z)$	polygonal approximation of an observation z
Q	set of cliques in a graph
R	compatibility coefficients
S	separability criterion
T	number of final weak classifiers in $H(\cdot)$
V	set of vertices in a graph
X	set of examples
Y	set of classes
Z	set of observations

Chapter 1
Introduction

Robots need to understand their environment in order to be able to perform different tasks within it. A robot's interface with the external world is usually composed of several sensors that gather data. How to understand and interpret this sensor information is one of the fundamental problems in mobile robots.

Building maps of the environment is a typical example of how a robot uses sensor information to interpret the world. Some of the most usual maps in mobile robotics represent the parts in the environment which are occupied by objects such as occupancy grid maps [3, 8]. The occupancy information enables the robots to navigate without collisions, or to localize itself inside the environment. To build an occupancy map, a mobile robot usually moves along a trajectory while gathering information with some proximity sensor able to detect obstacles. Creating maps of the environment is extensively studied in the research area of *Simultaneous Localization And Mapping* (SLAM) — or *Concurrent Mapping and Localization* (CML)— [2, 6, 9, 11, 13, 14].

Occupancy maps are useful models for localization and navigation, however, other robotic tasks require a higher level of knowledge about the environment. Spatial concepts like the distinction between different locations, connectivity between them, or global topologies can be acquired by a robot from its experience inside the environment as shown by Kuipers and colleagues in the *Spatial Semantic Hierarchy* and its extensions [1, 5, 10].

In the particular case of indoor environments, we can find typical divisions of space such as corridors, rooms, or doorways. These divisions can be grouped into categories represented by a *semantic term*. A semantic term relates a category to some functionality, objective situation, or possible affordance in the place it represents. For example, the term *corridor* refers to the places which include doorways leading to other rooms. Furthermore, the term *doorway* indicates the places representing transitions between two different rooms, or between a room and a corridor. The process of applying a semantic term to some division of the environment is also known as *semantic place labeling*.

For a lot of applications, robots can improve their service and human-robot communication if they are able to obtain a semantic categorization of the environment in

Ó.M. Mozos: Semantic Labeling of Places with Mobile Robots, STAR 61, pp. 1–5.
springerlink.com

which they work. As an example, a robot that possesses information about the type of places can be instructed to "open the door to the corridor." Moreover, semantic terms like *corridor* or *room* give the human user a more intuitive idea of the position of the robot in comparison to the raw 2D coordinates inside a map. In addition, the semantic place information about places can improve the performance of the robot in other tasks such as localization or exploration.

In this book, we consider the problem of semantically categorizing the different locations of indoor environments using a mobile robot. An example is given in Fig. 1.1. The image in Fig. 1.1(a) shows the occupancy grid map corresponding to the ground floor of building 52 at the University of Freiburg. In this map, only information about occupied and free space is given. However, some natural divisions can be extracted from this environment, as for example rooms, doorways and a corridor. The original occupancy map is then extended with information about these place categories as shown in Fig. 1.1(b).

As explained before, a robot gathers information about the environment through its sensors. Thus, the main question is how a mobile robot can recognize the different places of an indoor environment using sensor data. In this book we solve this problem using different classification techniques. In our specific case, a classifier is a function that maps sensor information into place categories. A mobile robot uses this function to determine the place in which it is located based on its current sensor observation. Before performing this mapping, the data coming from sensors is transformed into a feature vector which is fed into the classifier. This general approach comprises the basics of the different methods that will be presented along the book.

It is important to note that we will present methods that allow a robot to recognize places by categories. The difference between categories and instances is important. In our case, an instance represents a concrete place in the environment such as "room 1015 in building 79 at the University of Freiburg". A category, however, represents a set of instances. For example the category *room* would contain all indoor places that are likely to be a room. In general, categorization is considered a harder problem than instantiation because the classifier needs to be general enough to include all instances in one category, while at the same time needs to be specific enough to distinguish instances between different categories. In addition, when creating a classifier, the robot needs to find common features for all instances belonging to the same category.

All classifiers presented in this book are learned using supervision. Supervised learning is an approach which needs a set of already labeled examples to construct the mapping function. These examples are usually provided to the robot by an external agent or tutor. An example of a tutor can be a person showing the robot the different places in a house.

Opposite to supervised learning approaches are unsupervised techniques. A robot learns in a unsupervised fashion when there is no tutor indicating the different categories of places that can be found in the environment. In this case it is the robot itself who learns the different places according to some distinction measurement extracted from its sensor data [1, 4, 5, 16]. Typical unsupervised approaches use

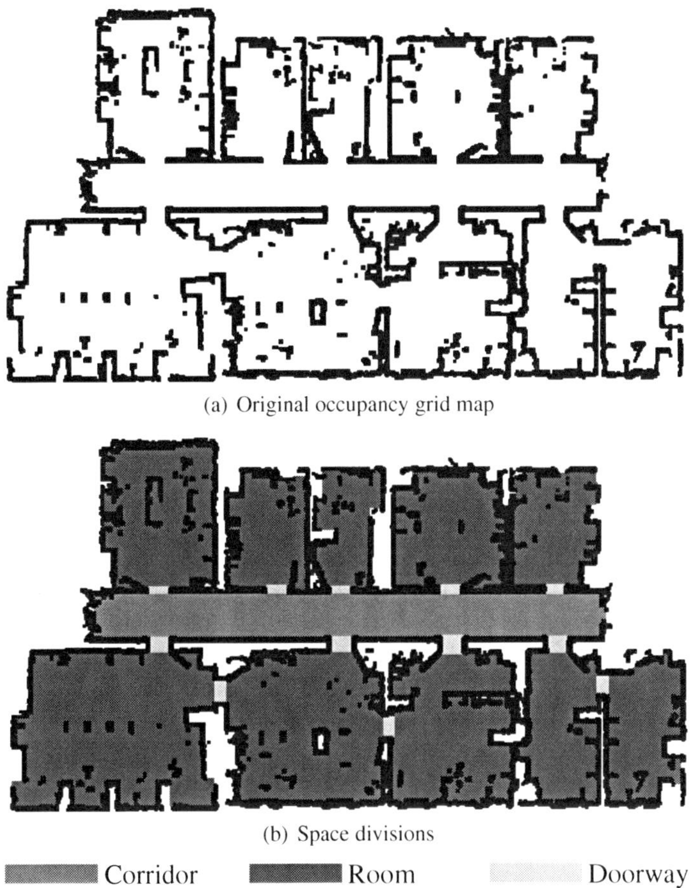

Fig. 1.1 Natural divisions of typical indoor environments. (**a**) shows the original occupancy grid map corresponding to the ground floor of building 52 at the University of Freiburg. (**b**) depicts some natural divisions extracted from the environment: rooms, doorways, and a corridor.

clustering [12] to separate the instances corresponding to each category. Unsupervised learning methods are outside the scope of this book.

The rest of the book is organized as follows. Chapter 2 gives a short introduction to supervised learning and presents the AdaBoost algorithm. For a more extensive introduction to supervised learning we refer the reader to [7, 12, 15].

In Chap. 3 we introduce the application of AdaBoost to assign semantic labels to different places in indoor environments using a mobile robot. The main idea is to classify each pose of a mobile robot into one of the semantic categories according to the laser range observation the robot gathered at that position. The classification is carried out using geometrical features extracted from the laser beams. Additionally, the boosting approach allows us to determine which are the most informative features used to recognize each place.

The method presented in Chap. 3 is extended in Chap. 4 to extract topological maps from indoor environments. The key idea is to apply the semantic classification to all possible poses of the robot in a map. In this way we obtain a complete categorization of the free space. Neighboring poses with similar classification are then grouped into regions which form the different nodes in the final topological map. Previous to the grouping, a smoothing method is applied that takes into account spatial dependencies between different labels.

The previous two methods cover the semantic classification of the different poses of a mobile robot, but they do not take into account the movement of the robot along a trajectory. When operating in indoor environments, the robot usually have a moderate velocity and a relatively continuous movement. That means, that observations obtained by a mobile robot at nearby poses are typically very similar. Based on this assumptions, Chap. 5 describes a method that takes into account previous classifications when classifying a new pose of a mobile robot along a trajectory. These spatial dependencies are modeled using a hidden Markov model. The robot used for the experiments in this chapter is equipped with a camera in addition to the laser sensor. The visual information is composed of objects extracted from the images. Using both sensors, the robot is able to distinguish a wider set of categories including kitchens, laboratories, offices, and seminar rooms.

The semantic labeling can be applied not only to improve the human-robot communication, but also to better carry out some other specific tasks for autonomous mobile robots. Chapter 6 presents the exploration of environments with a team of robots using place information. We will show how the semantic labeling of places can improve the distribution of the robots during the exploration. The main idea here is that corridors are better exploration targets as they lead to other rooms. In a second application, we will see how to accelerate the localization process of a single robot using the semantic classification of the different rooms.

Chapter 7 presents the semantic labeling of places as part of a high level conceptual representation of indoor environments called *multi-layer conceptual map*. This representation extends the semantic classification of places adding upper layers which include more complex conceptual terms, such as living rooms. The terms not only represent places but also objects such as TV sets or couches, and are used to create a human-friendly dialogue while interacting with people.

The above mentioned techniques are used to augment the information about environments with semantic terms. However, the AdaBoost-based classifiers can also be used to include semantic information in sensor data. Chapter 8 will show an approach to semantically label the beams of a laser scan. The main idea is to assign each beam the class of the object it hits. In this chapter, we restrict the classification to the labels *person* and *non-person*, although the method can be easily extended to use additional ones.

References

1. Beeson, P., Jong, N.K., Kuipers, B.: Towards autonomous topological place detection using the extended voronoi graph. In: Proceedings of the IEEE International Conference on Robotics and Automation (2005)
2. Dissanayake, G., Durrant-Whyte, H.F., Bailey, T.: A computationally efficient solution to the simultaneous localisation and map building (SLAM) problem. In: Proceedings of the IEEE International Conference on Robotics and Automation, San Francisco, CA, USA, April 2000, pp. 1009–1014 (2000)
3. Elfes, A.: Using occupancy grids for mobile robot perception and navigation. Computer 22(6), 46–57 (1989)
4. Fabrizi, E., Saffiotti, A.: Extracting topology-based maps from gridmaps. In: Proceedings of the IEEE International Conference on Robotics and Automation, San Francisco, CA, pp. 2972–2978 (2000)
5. Kuipers, B.: The spatial semantic hierarchy. Artificial Intelligence 119(1-2), 191–233 (2000)
6. Leonard, J.J., Durrant-Whyte, H.F.: Mobile robot localization by tracking geometric beacons. IEEE Transactions on Robotics and Automation 7(3), 376–382 (1991)
7. Mitchell, T.M.: Machine Learning. McGraw-Hill, New York (1997)
8. Moravec, H.P.: Sensor fusion in certainty grids for mobile robots. AI Magazine 9, 61–74 (1988)
9. Moutarlier, P., Chatila, R.: Stochastic multisensory data fusion for mobile robot location and environmental modelling. In: Fifth International Symposium of Robotics Research, Tokyo, Japan, pp. 85–94 (1989)
10. Remolina, E., Kuipers, B.: Towards a general theory of topological maps. Artificial Intelligence 152(1), 47–104 (2004)
11. Smith, R.C., Cheeseman, P.: On the representation and estimation of spatial uncertainty. The International Journal of Robotics Research 5(4), 56–68 (1986)
12. Theodoridis, S., Koutroumbas, K.: Pattern Recognition, 3rd edn. Academic Press, London (2006)
13. Thrun, S.: Robotic mapping: A survey. In: Lakemeyer, G., Nebel, B. (eds.) Exploring Artificial Intelligence in the New Millenium, pp. 1–34. Morgan Kaufmann, San Francisco (2002)
14. Thrun, S., Burgard, W., Fox, D.: Probabilistic Robotics. MIT Press, Cambridge (2005)
15. Witten, I.H., Frank, E.: Data Mining: Practical Machine Learning Tools and Techniques, 2nd edn. Morgan Kaufmann, San Francisco (2005)
16. Zivkovic, Z., Bakker, B., Kröse, B.: Hierarchical map building using visual landmarks and geometric constraints. In: Proceedings of the IEEE/RSJ International Conference on Intelligent Robots and Systems (2005)

Chapter 2
Supervised Learning

2.1 Introduction

In a supervised learning task we are interested in finding a function that maps a set of given examples into a set of classes or categories. This function, called *classifier*, will be used later to classify new examples, which in general are different from the given ones. The process of modeling the classifier is called *training*, and the algorithm used during the training is called *learning algorithm*. The initial given examples used by the learning algorithm are called *training examples*. The training examples are usually provided by an external agent to the robot also called *tutor*. Once the classifier is learned, it should be tested on new unseen examples to evaluate its performance. These new examples are called *test examples*.

Before using the examples (both training and test) in the classification process, they are usually transformed into a *feature vector*. A feature vector is a set of values that represent some characteristics of the classes we want to learn. In the more general case, each example x is represented by a set of L features in the form

$$x = \{f_1, \ldots, f_L\},$$

In the problems presented in this book each feature will be represented by a real value, i.e. $f_l \in \mathbb{R}$.

In the case of a binary classification, the classifier needs to distinguish between two classes only. The most intuitive way of separating two classes is by creating a hyperplane that separates the examples of both classes in the feature space. An example is shown in Fig. 2.1. In this figure two classes y_1 and y_2 are represented in a bidimensional feature space. In the training process, a hyperplane—a line in 2D—is calculated in a way that separates the feature space into two parts. Each division contains the examples of one class. In the classification step, a new example is represented in the feature space. The new example is assigned the class corresponding to the division of the space in which the example is located.

Ó.M. Mozos: Semantic Labeling of Places with Mobile Robots, STAR 61, pp. 7–13.
springerlink.com

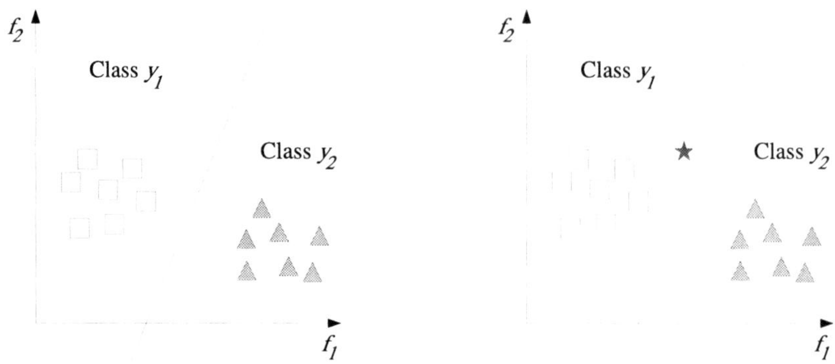

(a) Training step (b) Classification step

Fig. 2.1 (**a**) In the training step a line is calculated that separates the training examples of classes y_1 (*squares*) and y_2 (*triangles*). (**b**) In the classification step a new example (*star*) is classified according to the decision line. In this case it belongs to class y_1.

In general, the division of the feature space into different regions is done using functions that define different types of curves. These types of classifiers are called *discriminative*, since they just divide the feature space and discriminate the examples from the different classes.

The solution to the binary case can be used to learn one class only. To do this we divide the set of all possible examples in the world into two sets: a first set corresponding to the examples of the class we want to learn, and a second set containing the remaining examples. The examples of the learned class received the name *positive examples*, while the rest of the examples are called *negative examples*.

The learning of one class allows us to introduce some performance metrics for the classifiers. The evaluation of the performance is done using the set of test examples. These examples need to contain their corresponding class, since this value will be used in the metrics. After applying the learned classifier to the test set we can obtain the following values:

True positives (TP) Test examples whose original class was positive and are classified as positive.

True negatives (TN) Test examples whose original class was negative and are classified as negative.

False positives (FP) Test examples whose original class was negative but are classified as positive.

False negatives (FN) Test examples whose original class was positive but are classified as negative.

According to these values a good classifier will try to maximize the number of true classifications (TP and TN) while trying to minimize the number of false ones (FP and FN). These values are typically shown in a matrix form with two rows and two columns. This matrix is called *confusion matrix*. Moreover, these performance values are the basics for further metrics such as lift charts, receiver operating characteristic (ROC) curves, or recall–precision curves [16].

An important phenomenon that should be avoided when training a classifier is *overfitting*. Overfitting occurs when a classifier does very well in the training set but performs poorly in the test set. This situation usually occurs because the learned function fits too much to the training examples, leaving a small margin to detect new examples if they are slightly far away in the feature space. The capability of a classifier to detect new examples, even if they are slightly different from the training set is called *generalization*. Opposite to overfitting, generalization is a desired behavior in a classifier.

A classifier is also called a hypothesis, since it is a guess of the class to which an example belongs. The hypotheses that are only slightly better than a random guess are called *weak hypotheses*. Similarly, the classifiers that are much better than a random guess are called strong hypotheses.

Further details on supervised learning can be found in [9, 14, 16].

2.2 Boosting

Boosting is a general method which attempts to improve the accuracy of a given learning algorithm [4, 8, 12]. This approach has its roots in the probably approximately correct (PAC) framework [15].

Kearns and Valiant [7, 6] were the first to pose the question of whether a weak learning algorithm can be combined into an accurate strong learning algorithm. Later, Shapire [11] demonstrated that any weak learning algorithm can be efficiently transformed or boosted into a strong learning algorithm.

The underlying idea of boosting is to combine a set of T weak hypotheses

$$\{h_1, h_2, \ldots, h_T\}$$

to form a strong hypothesis H such that the performance of the strong hypothesis is better than the performance of each of the single weak hypothesis h_t in the form

$$H(x) = \sum_{t=1}^{T} \alpha_t h_t(x) , \tag{2.1}$$

where α_t denotes the weight of hypothesis h_t. Both α_t and the hypothesis h_t are to be learned within the boosting procedure. The resulting strong hypothesis H has the form of a weighted majority vote classifier.

The boosting algorithm proceeds as follows. The algorithm is provided with a set of labeled training examples $(x_1, y_1), \ldots, (x_N, y_N)$, where y_n is the label associated with instance x_n. On each round $t = 1, \ldots, T$, the boosting algorithm devises a weight distribution D_t over the set of examples, and selects a weak classifier h_t with low error ε_t with respect to D_t. Thus, the distribution D_t specifies the relative importance of each example for the current round. After T rounds, the booster must combine the weak classifiers into a strong one. The key idea is to alter the distribution over the training examples in a way that increases the weights of the harder elements, thus forcing the weak classifier to make less mistakes on these elements.

An important aspect related to boosting is overfitting. Large parts of the early literature explain that boosting would not overfit even when using a large number of rounds. However, simulations shown in [5, 10] indicate that data sets with high noise content could clearly show overfitting effects.

2.2.1 AdaBoost

The AdaBoost algorithm, introduced by Freund and Schapire [3], is one of the most popular boosting algorithms. Following the general idea of boosting, the AdaBoost algorithm takes as an input a training set of examples $(x_1, y_1), ..., (x_N, y_N)$, with y_n being the label associated with instance x_n. Since the algorithm is designed for binary classifications, the label for each example indicates whether it is positive $y_n = 1$ or negative $y_n = 0$. On each round $t = 1, \ldots, T$, AdaBoost calls a weak learning algorithm repeatedly to select a weak hypothesis h_t.

The AdaBoost algorithm differs from previous boosting algorithms [1, 2, 11] in that it needs no prior knowledge of the accuracies of the weak classifiers. During the boosting process, AdaBoost adapts to these accuracies and generates weighted majority hypotheses in which the weight of each weak hypothesis is a function of its accuracy.

The complete algorithm is described in Figure 2.2. In this algorithm, the distribution D indicates the importance of the examples at the beginning of the training process and it is controlled later by the iterative process. This distribution can be set initially as the uniform distribution so that $D_1(n) = 1/N$, meaning that all examples have the same importance at the beginning. On each round t, the algorithm maintains the normalized weight distribution $D_t(1), \ldots, D_t(N)$ over the training examples. The distribution D_t is fed to the weak learner which generates a classifier h_t that has a small error with respect to this distribution. The accuracy of the weak hypothesis h_t is measured by its error as

$$\varepsilon_t = \sum_{n=1}^{N} D_t(n) \left| h_t(x_n) - y_n \right| . \tag{2.2}$$

Notice that the error is measured with respect to the distribution D_t on which the weak learner was trained. In practice, the weak learning algorithm may be able to use the weights D_t on the training examples. Alternatively, when this is not possible, a subset of the training examples can be sampled according to D_t, and these resampled examples can be used to train the weak learner.

Using the new hypothesis h_t, the boosting algorithm generates the next weight vector D_{t+1}, and the process is repeated. After T iterations, the final strong hypothesis H is generated, combining the outputs of the T weak hypotheses using a weighted majority vote.

Freund and Schapire [3] proved that, for binary classification problems, the training error of the final hypothesis H generated by the AdaBoost algorithm is bounded by

$$\varepsilon \leq 2^T \prod_{t=1}^{T} \sqrt{\varepsilon_t(1-\varepsilon_t)} \leq \exp\left(-2\sum_{t=1}^{T}\gamma_t^2\right), \tag{2.3}$$

where $\varepsilon_t = 1/2 - \gamma_t$ is the error of weak hypothesis h_t. Since a hypothesis that makes an entirely random guess has error 0.5, γ_t measures the accuracy of the weak hypothesis h_t relative to a random guess. This bound shows that the final training error drops exponentially if each of the weak hypotheses is better than a random guess.

Since the accuracy of the AdaBoost algorithm depends on the fact that the weak classifiers are better than a random guess, an alternative way to stop the iterative process of Fig. 2.2 consists of testing whether the selected weak classifier h_t has a classification error better than 0.5. If this is not the case, then the loop should

- Input:
  - Set of N labeled examples $(x_1, y_1), \ldots, (x_N, y_N)$ with $y_n = 1$ if the example x_n is positive, and $y_n = 0$ if the example x_n is negative
  - Distribution D over the N examples
  - Weak learner
  - Integer T specifying the number of iterations
- Initialize the weight vector $D_1(n)$ for $i = 1, \ldots, N$.
- For $t = 1, \ldots, T$

  1. Normalize the weight distribution
  $$D_t(n) = \frac{D_t(n)}{\sum_{i=1}^{N} D_t(i)}$$
  2. Train weak learner using distribution D_t and get back a hypothesis $h_t : \mathrm{X} \rightarrow \{0, 1\}$.
  3. Calculate the error for h_t as
  $$\varepsilon_t = \sum_{n=1}^{N} D_t(n)\,|h_t(x_n) - y_n|$$
  4. Set
  $$\beta_t = \frac{\varepsilon_t}{(1-\varepsilon_t)}$$
  5. Set the new weights
  $$D_{t+1}(n) = D_t(n)\beta_t^{1-|h_t(x_n)-y_n|}$$
- The final strong hypothesis is given by:

$$H(x) = \begin{cases} 1 \text{ if } \sum_{t=1}^{T}\left(\log\frac{1}{\beta_t}\right)h_t(x) \geq \frac{1}{2}\sum_{t=1}^{T}\log\frac{1}{\beta_t} \\ 0 \text{ otherwise} \end{cases}$$

Fig. 2.2 The AdaBoost algorithm [3].

- Input:
  - Set of N labeled examples $(x_1, y_1), \ldots, (x_N, y_N)$ with $y_n = 1$ if the example x_n is positive, and $y_n = -1$ if the example x_n is negative
  - Weak learner
  - Integer T specifying the number of iterations
- Initialize the weights $D_1(n) = 1/N$ for $n = 1, \ldots, N$.
- For $t = 1, \ldots, T$
  1. Train the weak learner with distribution D_t and get back a hypothesis $h_t : X \to [0, 1]$.
  2. Choose $\alpha_t \in \mathbb{R}$.
  3. Update:
  $$D_{t+1}(n) = \frac{D_t(n)\exp(-\alpha_t y_n h_t(x_n))}{Z_t}$$
  where Z_t is a normalization factor so that D_{t+1} is a distribution
- The final strong hypothesis is given by:
$$H(x) = \text{sign}\left(\sum_{t=1}^{T} \alpha_t h_t(x)\right)$$

Fig. 2.3 Generalized version of the AdaBoost algorithm [13].

be finished, and the final strong classifier H constructed using the weak classifiers selected so far.

2.2.2 Generalized AdaBoost

An alternative version to the original AdaBoost algorithm was introduced in [13]. This version, called generalized AdaBoost, presents several improvements. First, the output of the weak hypotheses can have any real value inside the range $[0, 1]$ rather than only two values. Second, in this version of the algorithm the different α_t, which correspond to the weights of the final weak hypotheses, are left unspecified. The complete algorithm is shown in Figure 2.3.

In [13] a possible choice for the different weights α_t is presented

$$\alpha_t = \frac{1}{2}\ln\left(\frac{1 + r_t}{1 - r_t}\right), \tag{2.4}$$

where r_t is chosen at each iteration so that its absolute value $|r_t|$ is maximized according to

$$r_t = \sum_{n=1}^{N} D_t(n) y_n h_t(x_n). \tag{2.5}$$

Several versions of the generalized algorithm together with different comparisons are presented in [13].

References

1. Freund, Y.: Boosting a weak learning algorithm by majority. In: COLT: Proceedings of the Workshop on Computational Learning Theory. Morgan Kaufmann, San Francisco (1990)
2. Freund, Y.: Data filtering and distribution modeling algorithms for machine learning. PhD thesis, University of California at Santa Cruz (1993)
3. Freund, Y., Schapire, R.E.: A decision-theoretic generalization of on-line learning and an application to boosting. In: Proceedings of the European Conference on Computational Learning Theory, pp. 23–37 (1995)
4. Freund, Y., Schapire, R.E.: A short introduction to boosting. Journal of Japanase Society for Artificial Intelligence 14(5), 771–780 (1999)
5. Grove, A.J., Schuurmans, D.: Boosting in the limit: Maximizing the margin of learned ensembles. In: Proceedings of the National Conference on Artificial Intelligence, pp. 692–699 (1998)
6. Kearns, M., Valiant, L.G.: Cryptographic limitations on learning boolean formulae and finite automata. Journal of the ACM 41(1), 67–95 (1994)
7. Kearns, M.J., Valiant, L.G.: Learning boolean formulae or finite automata is as hard as factoring. Technical Report TR-14-88, Harvard University Aiken Computation Laboratory (August 1988)
8. Meir, R., Rätsch, G.: An introduction to boosting and leveraging. In: Mendelson, S., Smola, A.J. (eds.) Advanced Lectures on Machine Learning. LNCS (LNAI), vol. 2600, pp. 118–183. Springer, Heidelberg (2003)
9. Mitchell, T.M.: Machine Learning. McGraw-Hill, New York (1997)
10. Rätsch, G., Onoda, T., Müller, K.R.: Soft margins for adaboost. Machine Learning 42(3), 287–320 (2001)
11. Schapire, R.E.: The strength of weak learnability. Machine Learning 5, 197–227 (1990)
12. Schapire, R.E.: The boosting approach to machine learning: An overview. In: MSRI Workshop on Nonlinear Estimation and Classification (2001)
13. Schapire, R.E., Singer, Y.: Improved boosting algorithms using confidence-rated predictions. Machine Learning 37(3), 297–336 (1999)
14. Theodoridis, S., Koutroumbas, K.: Pattern Recognition, 3rd edn. Academic Press, London (2006)
15. Valiant, L.G.: A theory of the learnable. Communications of the ACM 27(11), 1134–1142 (1984)
16. Witten, I.H., Frank, E.: Data Mining: Practical Machine Learning Tools and Techniques, 2nd edn. Morgan Kaufmann, San Francisco (2005)

Chapter 3
Semantic Learning of Places from Range Data

3.1 Introduction

Building accurate maps of indoor environments is one of the typical problems in mobile robotics. In this task, a robot moves along a trajectory while gathering information with sensors. Typical maps represent the parts in the environment which are occupied by objects, as for example occupancy grid maps [5, 15]. The maps are then used for localization and navigation tasks. However, little work have been done to add semantic information to these maps. For a lot of applications, the service of robots can be improved if they are able to recognize places and differentiate them.

In this chapter, we address the problem of assigning semantic labels to locations of the environment using a mobile robot. Indoor environments, like the one depicted in Fig. 3.1, can typically be decomposed into areas with different functionalities, such as office rooms, corridors, entrance hallways, or doorways. Generally, each of these places has a different structure. For example, a corridor is usually longer than a room. Furthermore, rooms are typically smaller than hallways, and are also more cluttered than corridors or hallways.

The key idea of this chapter is to classify the position of the robot based on the current scan obtained from the range sensor. Fig. 3.2(a) shows an example range scan taken by a mobile robot in a corridor. Other examples for typical range scans obtained in an office environment are shown in Fig. 3.2(b).

The approach presented in this chapter uses the AdaBoost algorithm to boost simple geometrical features from the scans into a strong classifier. Each of this features alone is insufficient for a reliable categorization of places. The features are represented by a numerical value computed from the beams of a laser range scan as well as from a polygon representation of the covered area. Since AdaBoost provides only binary decisions, we determine the decision list with the best sequence of binary classifiers. Experimental results show that the resulting classification system can determine the type of the place with high classification rates. Moreover, results

Ó.M. Mozos: Semantic Labeling of Places with Mobile Robots, STAR 61, pp. 15–34.
springerlink.com

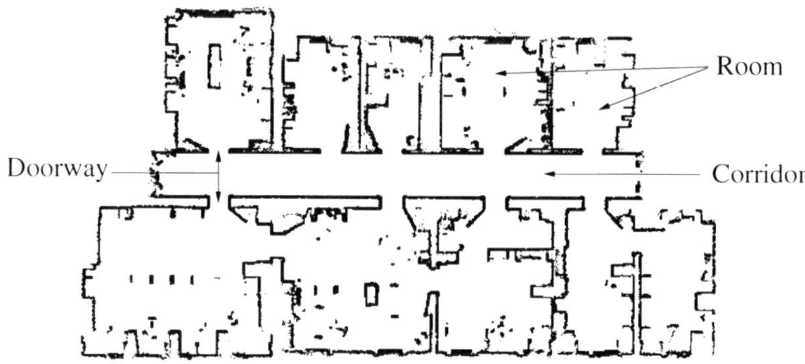

Fig. 3.1 Example environment containing rooms, doorways and a corridor.

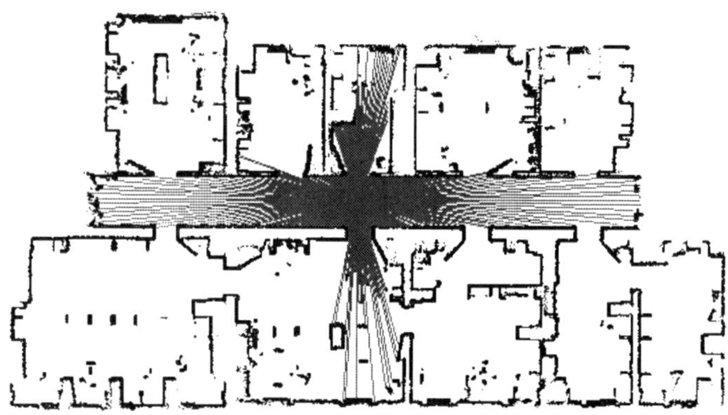

(a) Scan obtained in a corridor

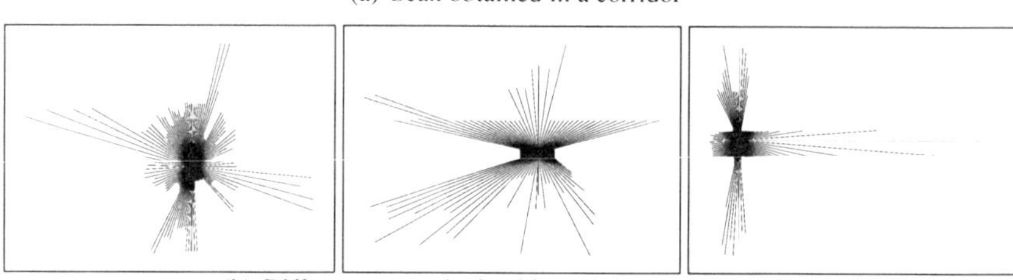

(b) Different scans obtained in an indoor environment

Fig. 3.2 (**a**) Range scan obtained by a mobile robot in a corridor. The image also shows the map of the environment where the scan was taken. The scan covers the complete 360° field of view of the robot. (**b**) Further scans recorded in a room, a doorway, and a corridor.

are presented illustrating that the final classifier can even be used in environments from which no training data was available.

Throughout this chapter we assume that the robot is equipped with a laser range scanner that covers 360° field of view around the robot as shown in Fig. 3.2(a).

However, common configurations on real mobile robots have only a laser covering 180° in front of the robot. We also present a solution for these cases.

The rest of the chapter is organized as follows. In the next section we describe the particular implementation of the AdaBoost algorithm for place labeling. Section 3.3 presents its extension to multiple classes. Sect. 3.4 introduces the features extracted from laser range scans. A solution to the problem of restricted field of view is given in Sect. 3.5. In Sect. 3.6, experimental results are presented. We discuss related work in Sect. 3.7. Finally, we conclude in Sect. 3.8.

3.2 Binary Classification Using AdaBoost

The AdaBoost algorithm, introduced in [6], is one of the most popular boosting algorithms (cf. Sect. 2.2). AdaBoost is designed to find a binary classifier that discriminates between positive and negative examples. The input to the learning algorithm is a set of training examples $(x_n, y_n), n = 1, \ldots, N$, where each x_n is an example and $y_n \in \{1, 0\}$ is a label indicating whether x_n is a positive example ($y_n = 1$) or a negative one ($y_n = 0$). In addition, the examples are weighted according to a distribution D_t indicating their relative importance. In a set of rounds $t = 1, \ldots, T$, AdaBoost improves the classification performance of a simple learning algorithm by boosting a collection of weak classifiers h_t to a stronger one. Each weak classifier h_t returns a value $\in \{1, 0\}$ indicating whether the example is positive or negative respectively. A condition to each weak classifier is that its classification rate must be better than a random guess. On each round t, the examples are re-weighted in order to increase the importance of those which were incorrectly classified by the previous selected weak classifier. The final strong classifier takes the form of a weighted combination of weak classifiers followed by a threshold. Large weights are assigned to good classification functions whereas poor functions have small weights.

Throughout this work we apply the variant of the AdaBoost algorithm presented by Viola and Jones [28]. This implementation restricts the weak classifiers to depend on single-valued features f_j only. Each weak classifier has the form

$$h_j(x) = \begin{cases} 1 & \text{if } p_j f_j(x) < p_j \theta_j \\ 0 & \text{otherwise}, \end{cases} \tag{3.1}$$

where θ_j is a threshold, and p_j is either -1 or 1 and represents the direction of the inequality. For each weak classifier h_j, the algorithm determines the optimal values for θ_j and p_j such that the number of misclassified training examples is minimized as

$$(p_j, \theta_j) = \underset{(p_i, \theta_i)}{\operatorname{argmin}} \sum_{n=1}^{N} D_t(n) \left| h_i(x_n) - y_n \right| . \tag{3.2}$$

The final algorithm is given in Fig. 3.3.

- Input:
  - Set of N labeled examples $(x_1, y_1), \ldots, (x_N, y_N)$
    with $y_n = 1$ if the example x_n is positive,
    and $y_n = 0$ if the example x_n is negative
  - Integer T specifying the number of iterations
- Let l be the number of positive examples and m the number of negative examples. Initialize weights $D_1(n) = \frac{1}{2l}$ when $y_n = 1$, and $\frac{1}{2m}$ when $y_n = 0$.
- For $t = 1, \ldots, T$
  1. Normalize the weight distribution
  $$D_t(n) = \frac{D_t(n)}{\sum_{i=1}^{N} D_t(i)}$$
  2. For each feature f_j train weak classifier h_j using distribution D_t.
  3. For each h_j, calculate the error ε_j as
  $$\varepsilon_j = \sum_{n=1}^{N} D_t(n) \left| h_j(x_n) - y_n \right|$$
  4. Choose the classifier h_j with lowest error ε_j and set $(h_t, \varepsilon_t) = (h_j, \varepsilon_j)$.
  5. Set
  $$\beta_t = \frac{\varepsilon_t}{(1 - \varepsilon_t)}$$
  6. Set the new weights
  $$D_{t+1}(n) = D_t(n)\beta_t^{1 - |h_t(x_n) - y_n|}$$
- The final strong classifier is given by:
$$H(x) = \begin{cases} 1 & \text{if } \sum_{t=1}^{T} \left(\log \frac{1}{\beta_t}\right) h_t(x) \geq \frac{1}{2} \sum_{t=1}^{T} \log \frac{1}{\beta_t} \\ 0 & \text{otherwise} \end{cases}$$

Fig. 3.3 The AdaBoost algorithm for place categorization using laser-based features.

3.3 Classification of Multiple Classes

The previous AdaBoost algorithm was designed for binary classification problems. However, to label places in the environment we need the ability to handle multiple classes. A way of constructing a multi-class classifier is to arrange several binary classifiers into a decision list. Each element of such a list represents one binary classifier, which determines if an example belongs to one specific class. If the classifier returns a positive result, the example is assumed to be correctly classified. Otherwise the example is passed to the next classifier in the list. The structure of such a decision list is shown in Fig. 3.4.

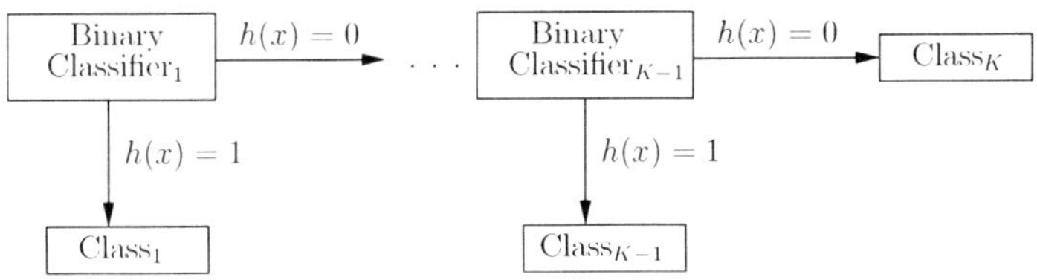

Fig. 3.4 A decision list for *K* classes using binary classifiers.

One important question in the context of a sequential classifier is the order in which the individual binary classifiers are arranged. This order can have a major influence on the overall classification performance, because the individual classifiers typically have different accuracies. In general, the problem of finding the optimal order of binary classifiers that minimizes the classification error is NP-hard. However, in our application we deal with a small number of classes, therefore we can enumerate all potential permutations and choose the optimal sequence.

To train each of the binary classifiers in the decision list we follow the *one-against-all* approach. In this technique, each classifier is trained using the examples of a concrete class as the positive set, and the examples corresponding to all other class as the negative set (cf. Sect. 2.1). In this way, each classifier is trained to work as a detector for one specific class.

3.4 Simple Features from Sensor Range Data

In the previous section we described the key principles of the AdaBoost algorithm for boosting simple features to strong classifiers. It remains to describe the features of the range scans used in the system. We assume that the mobile robot is equipped with a 360° field of view range sensor. Each observation z contains a set of beams in the form

$$z = \{b_0, ..., b_{M-1}\} .$$

Each beam b_i consists of a tuple (ρ_m, d_m) where ρ_m is the angle of the beam relative to the robot and d_m is the length of the beam.

The method for place classification is based on simple geometrical features extracted from the range scans observations z. We call them *simple* because they are single-valued features. All features are rotational invariant to make the classification of a pose dependent only on the (x,y)-position of the robot and not on its orientation. Most of the features are standard geometrical features often used in shape analysis [8, 9, 13, 18, 22].

We define a feature f as a function that takes as argument one observation and returns a real value: $f : \mathrm{Z} \rightarrow \mathbb{R}$, where Z is the set of all possible observations. Two sets of simple features are calculated for each observation. The first set A is calculated using the raw beams in z (cf. Fig. 3.2(b)). The following is a list of the single-valued features pertaining to this set:

1. The average difference between the length of consecutive beams.
2. The standard deviation of the difference between the length of consecutive beams.
3. The average difference between the length of consecutive beams considering different max-range values.
4. The standard deviation of the difference between the length of consecutive beams considering different max-range values.
5. The average beam length.
6. The standard deviation of the beam length.
7. The number of gaps in the scan. Two consecutive beams build a gap if their difference is greater than a given threshold. Different features are used for different threshold values.
8. The number of beams lying on lines that are extracted from the range scan [23].
9. The Euclidean distance between the two points corresponding to two consecutive global minima.
10. The angular distance between the two points corresponding to two consecutive global minima.

In this list, a max-range value indicates the maximum distance that a laser beam can reach. Over this distance the beam always provides a max-range reading.

The second set B of features is calculated from a polygonal approximation $\mathrm{Pol}(z)$ of the area covered by z. The vertices of the closed polygon $\mathrm{Pol}(z)$ correspond to the coordinates of the end-points of each beam b_m of z relative to the robot in the form

$$\mathrm{Pol}(z) = \{(d_m \cos\rho_m, d_m \sin\rho_m) \mid m = 0, \ldots, M-1\} \,. \tag{3.3}$$

As an example, the polygonal representations of the laser range scans depicted in Fig. 3.2(b) are shown in Fig. 3.5. The list of features corresponding to the set B is as follows:

1. Area of $\mathrm{Pol}(z)$.
2. Perimeter of $\mathrm{Pol}(z)$.
3. Area of $\mathrm{Pol}(z)$ divided its perimeter.
4. Mean distance between the centroid and the shape boundary.
5. Standard deviation of the distances between the centroid and the shape boundary.
6. Similarity invariant descriptors based on the Fourier transformation. We use the first 200 descriptors.
7. Major axis Ma of the ellipse that approximates $\mathrm{Pol}(z)$ using the first two Fourier coefficients.
8. Minor axis Mi of the ellipse that approximate $\mathrm{Pol}(z)$ using the first two Fourier coefficients.
9. Ma/Mi.
10. Seven invariants moments of $\mathrm{Pol}(z)$.
11. The normalized feature of compactness of $\mathrm{Pol}(z)$.

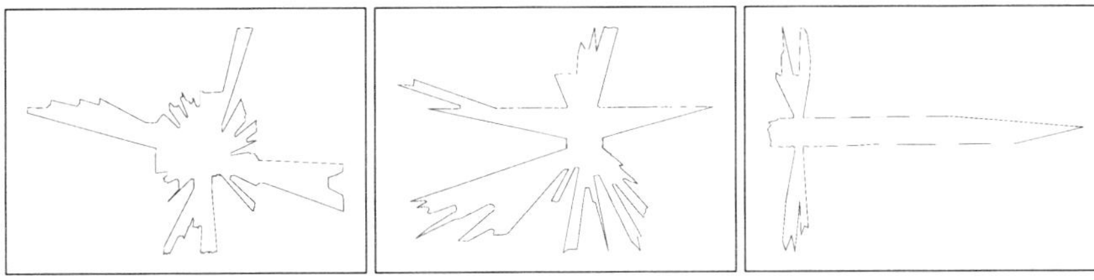

Fig. 3.5 Example of polygon approximations of scans recorded in a room, a doorway, and a corridor.

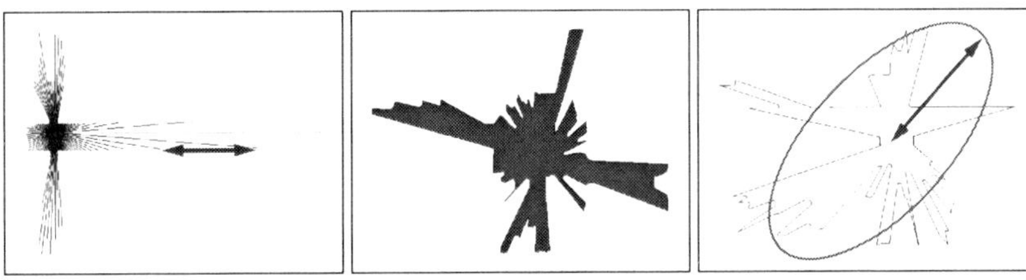

Fig. 3.6 Examples of features generated from laser data, namely the average distance between two consecutive beams, the perimeter of the area covered by a scan, and the mayor axis of the ellipse that approximates the polygon described by the scan. The laser beams cover a 360° field of view.

12. The normalized feature of eccentricity of $\mathrm{Pol}(z)$.
13. The form factor of $\mathrm{Pol}(z)$.

We will refer to the features in each set as A.i, and B.i, with the i-index identifying the position of the feature in the corresponding list. Fig. 3.6 shows graphically some of the features from the A and B sets. More specifically, the features are the average distance between two consecutive beams (A.1), the area covered by a scan (B.1), and the major axis of the ellipse that approximates the polygon described by the scan (B.7). The complete feature lists, together with their mathematical definition can be found in Appx. A and Appx. B.

Finally, each laser observation z is transformed into a feature vector in the form

$$x = \tau(z) = \{f_1, \ldots, f_L\} ,$$

with $f_j \in \mathrm{A} \bigcup \mathrm{B}$. Each training example for the AdaBoost algorithm consists of one observation represented by its feature vector x, together with its classification y. Thus, the set of training examples is given by

$$\{(x_n, y_n) \mid y_n \in \mathrm{Y}\} ,$$

where $\mathrm{Y} = \{\text{corridor}, \text{room}, \ldots\}$ is the set of classes corresponding to the places we want to recognize. Throughout this chapter we assume that the classification of the training examples is given in advance. In practice this can be achieved by manually labeling places in the map or by instructing the robot while it is exploring its environment.

3.5 Feature Extraction with Restricted Field of View[1]

As mentioned in the previous section, the simple features are based on laser observations covering 360° field of view as shown in Fig. 3.7(a). However, common configurations on real mobile robots have only a laser covering 180° in front of the robot. This situation is depicted in Fig. 3.7(b). In these last cases we propose to maintain a local map around the robot when classifying a pose of the robot during a trajectory. This local map can be updated during the movements of the robot, and then used to simulate the rear laser beams. An example of a local map is shown in Fig. 3.7(c).

In our case, we maintain a sparse local map around the robot. This map contains the endpoints of the previous laser beams that hit some object around the robot. We simulate the rear beams using this sparse map. For each simulated beam that does not hit any object in the sparse map, we calculate its value using an interpolation between the values of their (known) neighboring beams at both sides.

3.6 Experimental Results

The complete approach described in this chapter has been implemented and tested on a real robot as well as in simulation using the Carnegie Mellon Robot Navigation Toolkit (CARMEN) [4, 14]. The robots used to carry out the experiments were an ActivMedia Pioneer 2-DX8 equipped with two SICK laser range finders, and a PowerBot robot equipped only with a front laser. Both robots are shown in Fig. 3.8. The goal of the experiments is to demonstrate that the simple features can be boosted to a robust classifier of places. Additionally, we analyze whether the resulting classifier can be used to classify places in environments for which no training data were available. We first describe the results obtained with the sequential version of AdaBoost. In the next experiment we analyze how a mobile robot can utilize the resulting classifier along a trajectory. Furthermore, we present an experiment illustrating that a classifier can be applied to robustly classify places in a completely new environment.

One important parameter of the AdaBoost algorithm is the number of weak classifiers T used to form the final strong classifier. In total we formulated more than 300 simple features, each of them with two free parameters, which are determined in the learning phase according to (3.2). AdaBoost even uses features multiple times with different parameters. Thus, much more than the initial sets of simple features are available to form the strong classifier. We performed several experiments with different numbers of weak classifiers and analyzed the classification error. Throughout the experiments, we found that 100 weak classifiers provide the best trade-off

[1] The work presented in this section originated from a collaboration with Patric Jensfelt.

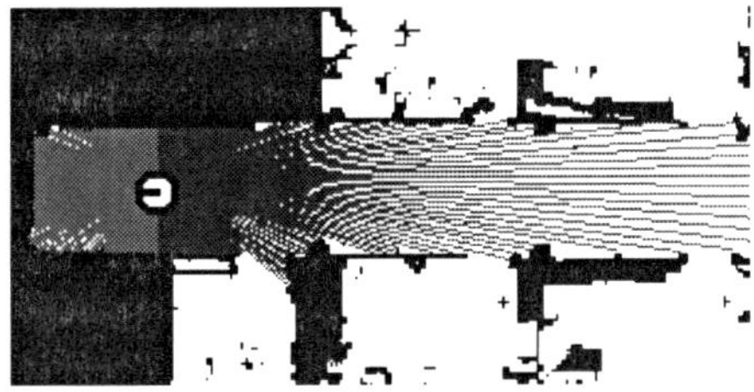

(a) 360° field of view

(b) 180° field of view

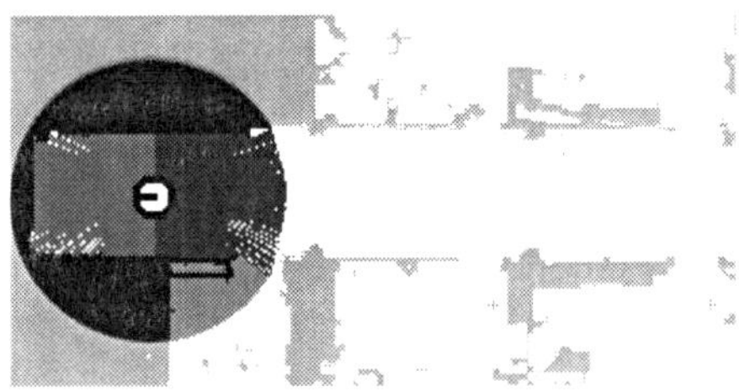

(c) Local map around the robot

Fig. 3.7 (**a**) Laser observation covering 360° field of view. (**b**) Laser observation covering only 180° in front of the robot. (**c**) Rear beams are simulated using the occupancy information contained inside the local map (*shaded area*).

between the error rate of the classifier and the computational cost of the algorithm. Therefore we used this value in all the experiments presented in this chapter.

3.6.1 Results Using Decision Lists

The following experiments illustrate that the approach presented in this chapter is well-suited to classify places in indoor environments according to a single laser range scan. In all the experiments, we first obtained an occupancy grip map representation of the environment. This map was then divided into two parts. One part was used as the training set, and the other part was used as the test set. For obtaining the training examples a simulated robot was situated in each free cell of the occupancy map using the CARMEN toolkit. At each position, an observation covering 360° field of view was simulated. The set of training observations were manually labeled according to the place divisions in the training part of the map. In a second step, the simulated robot took a set of observations in the test part of the map. These test examples were used for obtaining the classification rates.

The first experiment was performed using data from the office environment in building 79 at the University of Freiburg. This environment contains three different types of places, namely rooms, doorways, and a corridor. The left half of the environment contains the positions in which the robot is located to obtain the training examples, while the right half of the environment was used as a test set as shown in Fig. 3.9(a). In this experiment, we used a probabilistic sequential classifier as introduced in Sect. 3.3. In this particular case, each binary classifier identifies one place of the environment, i.e. room, door or corridor. Because we only have three classes, we tried all the possible combinations of binary classifiers for the decision list. We used the sequential classifier room-doorway which gives the best results and correctly classifies 93.94% of the test examples. The classification results are depicted as colored/gray areas in Fig. 3.9(b).

Table 3.1 contains the classification results of the other five potential sequential classifiers applied to the office environment of building 79. As can be seen in this table, the worst configurations are those in which the doorway classifier is in the first place. The corridor-doorway classifier also fails to perform well. The best configurations are corridor-room, room-doorway, and room-corridor.

A similar experiment was carried out in the office environment of building 52 at the University of Freiburg. Results are shown in Fig. 3.10. For this environment, the best sequence of binary classifiers was room-corridor with a classification rate of 92.10%. The classification rates for the different decision list configurations are shown in Table 3.2.

A third experiment was performed using a map containing four different classes, namely rooms, corridors, doorways, and hallways. The occupancy map with the training and test sets is depicted in Fig. 3.11(a). This map was constructed using the hallway of building 101 at the University of Freiburg and adding some rooms from other buildings to complete the four different classes. In this environment, we

(a) ActivMedia Pioneer (b) PowerBot

Fig. 3.8 (**a**) ActivMedia Pioneer 2-DX8 equipped with two SICK laser range finders. (**b**) PowerBot robot equipped with a front laser.

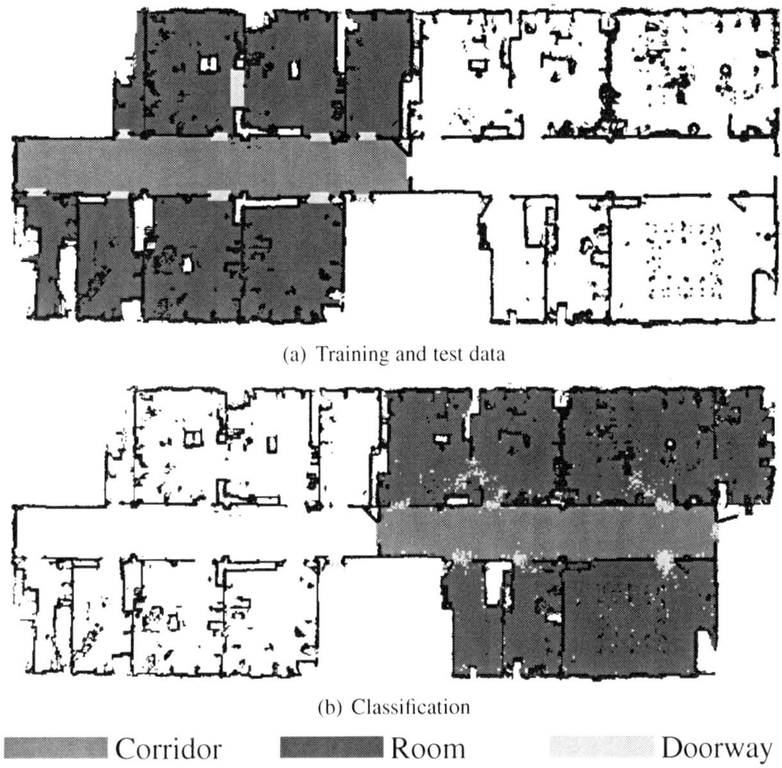

(a) Training and test data

(b) Classification

Fig. 3.9 (**a**) Training data (*colored*) and test data (*white*) from the office environment in building 79 at the University of Freiburg. (**b**) Classification results after applying the decision list room-doorway to the test set.

Table 3.1 Classification rates for the 6 sequential configurations in building 79.

Classifier Sequence	Correct Classifications %
room-doorway	93.94
room-corridor	93.31
corridor-room	93.16
doorway-corridor	80.68
doorway-room	80.49
corridor-doorway	80.10

obtained the best classification using the decision list corridor-hallway-doorway with a success rate of 89.52%. The resulting classification is depicted in Fig. 3.11(b). In addition, Table 3.3 shows the classification results for all possible sequential combinations of the four classifiers.

Finally, Table 3.4 contains the error rates of the individual binary classifiers on the training data of Fig. 3.11, which contains four classes. The error-rates differ between

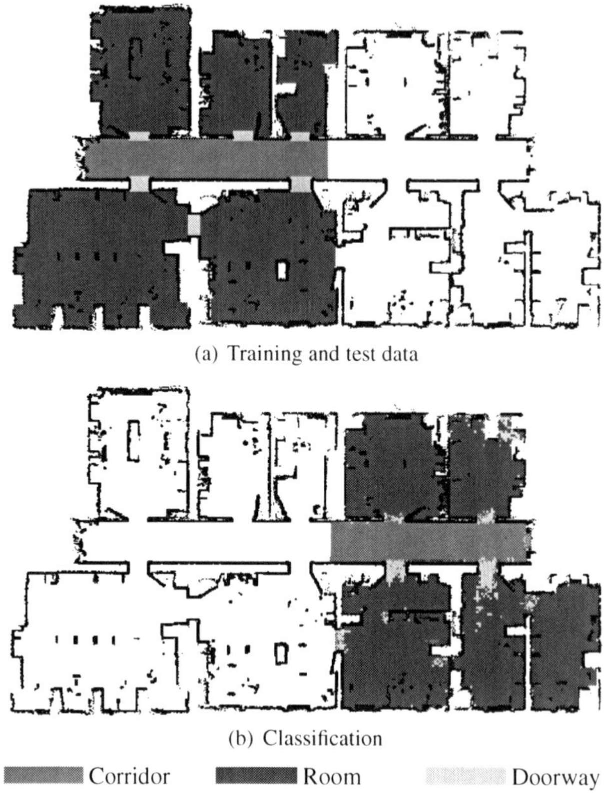

Fig. 3.10 (**a**) Training data (*colored*) and test data (*white*) from the office environment in building 52 at the University of Freiburg. (**b**) Classification results after applying the decision list room-corridor to the test set.

Table 3.2 Classification rates for the 6 sequential configurations in building 52.

Classifier Sequence	Correct Classifications %
room-corridor	92.10
corridor-room	91.57
doorway-corridor	91.13
corridor-doorway	91.03
room-doorway	90.94
doorway-room	90.30

0.7% and 1.5%. The binary doorway classifier yields the highest error. We believe that this is due to several reasons. First, a doorway typically is a very small area so that only a few training examples are available. Furthermore, if a robot stands in a doorway the scan typically covers nearby rooms or corridors, which make it hard to distinguish the doorway from such places.

Fig. 3.11 (**a**) Training data (*colored*) and test data (*white*) from the environment with four classes. (**b**) Classification results after applying the decision list corridor-hallway-doorway to the test set.

3.6.2 Transferring the Classifiers to New Environments

The next experiment is designed to analyze whether a classifier learned in a particular environment can be used to successfully classify the places of a new unseen environment. To carry out this experiment we used the sequential AdaBoost classifier room-corridor trained in the map shown in Fig. 3.10(a). This classifier was then evaluated on scans simulated given the map of the Intel Research Lab in Seattle shown in Fig. 3.12(a). The resulting classification is depicted in Fig. 3.12(b). In this environment, the classification rate was of 82.23%. This indicates that our algorithm yields good generalizations and can also be applied to correctly label places of so far unknown environments. Note that a success rate of 82.23% is quite high for this environment, since even humans typically do not consistently/correctly classify the places in this environment.

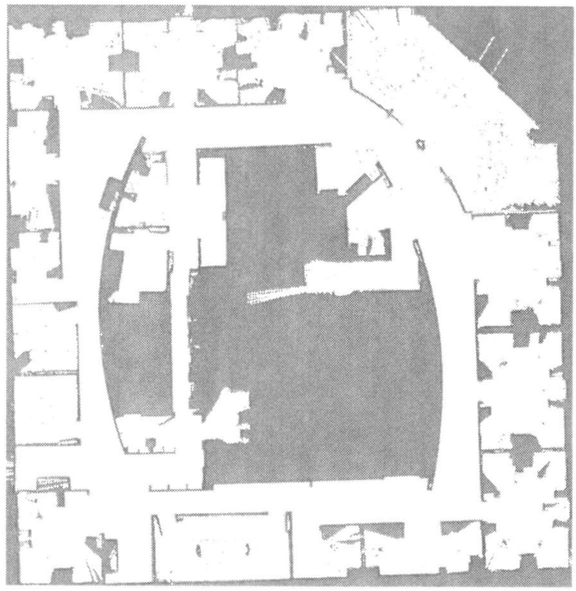

(a) Test map

(b) Classification

Corridor Room Doorway

Fig. 3.12 (**a**) Map of the Intel Research Lab in Seattle. (**b**) Classification results obtained by applying the classifier learned for the environment depicted in Fig. 3.10(a). The fact that 82.23% of all places could be correctly classified illustrates that resulting classifiers can be applied to so far unknown environments.

Table 3.3 Classification rates for the 24 sequential configurations in the four classes environment.

Classifier Sequence	Correct Classifications %
corridor-hallway-doorway	89.52
corridor-room-hallway	89.41
corridor-hallway-room	89.36
room-corridor-hallway	89.29
hallway-room-doorway	88.95
room-hallway-doorway	88.78
hallway-corridor-doorway	88.69
hallway-doorway-corridor	88.68
hallway-doorway-room	88.59
hallway-corridor-room	88.53
hallway-room-corridor	88.53
room-hallway-corridor	88.36
corridor-room-doorway	86.88
room-corridor-doorway	86.81
room-doorway-corridor	86.80
corridor-doorway-room	86.60
doorway-room-corridor	86.59
doorway-corridor-room	86.57
corridor-doorway-hallway	85.82
doorway-corridor-hallway	85.74
room-doorway-hallway	84.98
doorway-hallway-room	84.76
doorway-hallway-corridor	84.75
doorway-room-hallway	84.68

Table 3.4 Error in the training data for the individual binary classifiers learned from the map depicted in Fig. 3.11 containing four classes.

Binary Classifier	Training error %
corridor	0.7
hallway	0.7
room	1.4
doorway	1.5

3.6.3 Place Recognition with a Moving Robot

The goal of the following experiment is to show how a mobile robot can use our classifier to detect the different places along a trajectory. We use the best classifier for our office building 79, which was room-doorway according to Table 3.1, to classify the current pose of a mobile robot. A Pioneer2-DX8 robot, as the one shown in Fig. 3.8(a), was steered through the corridor, different rooms, and several doorways. While the robot was moving we logged its trajectory and the classifications obtained for the different range scans. The result is depicted in Fig. 3.13. Again, the

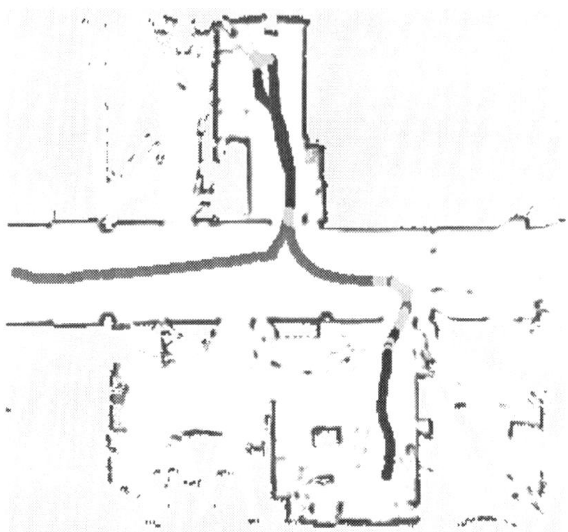

Fig. 3.13 Classification results obtained with the mobile robot of Fig. 3.8(a) moving through our office environment. Colors/gray levels in the image indicate the classification of the corresponding places on the trajectory.

different colors/gray levels of the points on the trajectory indicate the classification of the corresponding scan. As can be seen, the robot reliably identifies the type of the place. Only a few places show wrong classifications. These failures are mostly caused by clutter in the environment which make the sequential room-doorway classifier believe that the current place is a doorway. The process of extracting features from one observation and classifying it is fast enough to be carried out in real-time. Note that the geometrical features are very fast to calculate, and the final classifier is composed of a weighted sum and a comparison.

3.6.4 Classification of Trajectories Using Sensors with Restricted Field of View

In this experiment we present the results of applying the previous classification methods when the laser range scan has a restricted field of view. For this experiment we used a PowerBot robot equipped with a front laser covering 180° in front of the robot as shown in Fig. 3.8(b). We first steered the robot from the right end to the left end of the 6th floor of the CAS building at KTH. The trajectory is shown in Fig. 3.14(a). The data recorded on this floor was used to train the AdaBoost classifier. We then classified a trajectory on the 7th floor in the same building. We started the trajectory in an opposite direction (left end to right end). The rear beams were simulated using a local map as explained in Sect. 3.5. The resulting classification rate of 84.4% is depicted in Fig. 3.14(b). As the results indicate, restricting the field of view decreases the classification rate, however it maintains acceptable levels.

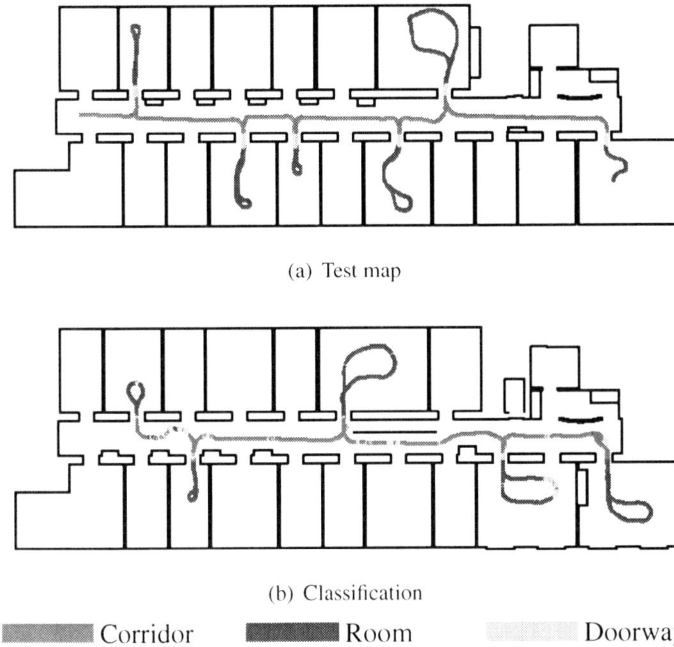

(a) Test map

(b) Classification

Fig. 3.14 (**a**) Training trajectory on the 6th floor of the CAS building at KTH. (**b**) Labeling of the trajectory on the 7th floor using only a front laser. The classification rate was of 84.4%. The map shown is for informative purposes only and does not represent exactly the environment in which the experiments were carried out, since the furniture is not showed.

3.6.5 Selected Features

Finally, we analyze the importance of the individual features in the final strong classifier. Table 3.5 lists the seven best features for each binary classifier with the left-most feature the most important. The individual classifiers were trained in the map of Fig. 3.11(a) and applied to the corresponding test set in Fig. 3.11(b). Note that often identical features occur. These features differ in their parameters (p_j, θ_j), and the weight α_j of the corresponding final weak classifier. As the table shows, several features like the average difference between consecutive beams (A.1) appears to be quite important. Furthermore, the number of gaps (A.7), which represents how

Table 3.5 Best five features for each binary classifier.

binary classifier	seven best features
corridor	A.7, A.1, B.7, B.6, B.6, A.1, A.1
room	B.2, A.1, B.4, B.6, B.7, A.7, B.5
doorway	A.9, A.1, A.10, A.5, A.2, B.6, A.1
hallway	B.1, A.1, A.9, B.1, B.12, B.6, A.1

cluttered the environment is, appears quite often. Whereas feature B.1, which corresponds to the area of the polygon, is most important for the detection of hallways, the feature A.9, which measures the distance between the smallest local minimum in the range scan, has the highest weight in the classifier for doorways.

3.7 Related Work

In the past, several authors considered the problem of adding semantic information to places. For example, Koenig and Simmons [11] use a pre-programmed routine to detect doorways from range data. Additionally, Althaus and Christensen [1] use line features to detect corridors, and free spaces between segments to detect doorways. The parameters of these models are set by hand. Compared to these approaches, the algorithm presented in this chapter does not require pre-defined parameters for extracting high-level features. Instead, it uses boosting to automatically learn the best features for identifying places.

In their work, Buschka and Saffiotti [3] describe a virtual sensor based on mathematical morphology that is able to identify rooms from range data. This technique requires the complete map to detect the rooms. Opposite to this method is the approach presented in this chapter, which allows a mobile robot to determine the place using the current observation only.

Oore et al. [17] present an interesting approach to learn parameters for the observation models from the robot position inside a map measured in (x, y) coordinates. This is actually the inverse problem to the one presented in this book. The main idea in [17] was to improve the localization of the robot when using noisy sensors as sonars.

Torralba et al. [26] use hidden Markov models for learning places using image data. This approach was able to distinguish concrete places along a trajectory. However, in [26] the authors apply instance classification instead of categorization.

Additionally, Kuipers [12] detect distinctive states in the map in an unsupervised manner. The states are used as places to create a topological map. However, the places in this map do not represent concrete spaces such as rooms or corridors, although they can have some relation to them.

Boosting has been used to identify objects using different features. Perhaps one of the most famous applications of AdaBoost is the fast recognition of faces in images by Viola and Jones [28]. Our algorithm is similar to the one presented in [28], since we also create simple features for the classification. However, the problem to be solved is totally different, because we classify locations in indoor environments using 2D range data. Boosting simple features is also used by Persson et al. [19] to create a virtual sensor for the detection of buildings outdoors, and by Treptow et al. [27] to track soccer balls.

The approach of this chapter was first introduced in [16]. Several posterior works have used similar ideas. For instance, Friedman et al. [7] present an approach for the classification of places using Voronoi random fields. This work also uses simple features that are selected using boosting as characteristics for the nodes in a Markov random field. The paper by Pronobis et al. [20] shows an approach to classify the different places of an indoor environment using vision. An extension to this work has been recently introduced in [21], in which the classification of places is done using an additional laser sensor together with the set of features presented in this chapter. Moreover, Topp and Christensen [25] use a similar idea of describing regions with simple geometrical features extracted from laser readings. The work by Brunskill et al. [2] presents an online method for generating topological maps from raw sensor information based on spectral clustering. In [2], the laser observations are represented by the set of features presented in this chapter. Finally, Sousa et al. [24] apply the same set of features for classifying places in indoor environments. Instead of AdaBoost, they use a support vector machine as classifier.

The semantic labeling of places has also been used as the base for other high level tasks. For example, the approach by Kersting et al. [10] shows that the semantic classification of places can be used to learn navigation policies using relational Markov decision processes.

3.8 Conclusion

In this chapter we presented a supervised approach to classify different places in the environment into semantic classes such as rooms, corridors, doorways, and hallways. The described technique uses simple geometric features extracted from a single laser range scan and applies the AdaBoost algorithm to learn a strong classifier. To distinguish between more than two classes we use a sequence of binary classifiers arranged in a decision list. Experiments carried out on a real robot as well as in simulation illustrate that our technique is well-suited to classify places in different environments even without training the classifier for each environment.

References

1. Althaus, P., Christensen, H.I.: Behaviour coordination in structured environments. Advanced Robotics 17(7), 657–674 (2003)
2. Brunskill, E., Kollar, T., Roy, N.: Topological mapping using spectral clustering and classification. In: Proceedings of the IEEE/RSJ Conference on Intelligent Robots and Systems, San Diego, CA, USA (October 2007)
3. Buschka, P., Saffiotti, A.: A virtual sensor for room detection. In: Proceedings of the IEEE/RSJ International Conference on Intelligent Robots and Systems, pp. 637–642 (2002)
4. CARMEN. Carnegie Mellon Robot Navigation Toolkit, `http://carmen.sourceforge.net/`
5. Elfes, A.: Using occupancy grids for mobile robot perception and navigation. Computer 22(6), 46–57 (1989)
6. Freund, Y., Schapire, R.E.: A decision-theoretic generalization of on-line learning and an application to boosting. In: Proceedings of the European Conference on Computational Learning Theory, pp. 23–37 (1995)
7. Friedman, S., Pasula, H., Fox, D.: Voronoi random fields: Extracting the topological structure of indoor environments via place labeling. In: Proceedings of the International Joint Conference on Artificial Intelligence, Hyderabad, India (2007)
8. Gonzalez, R.C., Wintz, P.A.: Digital Image Processing. Addison-Wesley Publishing Inc., Reading (1987)

9. Haralick, R.M., Shapiro, L.G.: Computer and Robot Vision. Addison-Wesley Publishing Inc., Reading (1992)
10. Kersting, K., Plagemann, C., Cocora, A., Burgard, W., De Raedt, L.: Learning to transfer optimal navigation policies. Advanced Robotics. Special Issue on Imitative Robots 21(9) (September 2007)
11. Koenig, S., Simmons, R.G.: Xavier: A robot navigation architecture based on partially observable markov decision process models. In: Kortenkamp, D., Bonasso, R., Murphy, R. (eds.) Artificial Intelligence Based Mobile Robotics: Case Studies of Successful Robot Systems, pp. 91–122. MIT Press, Cambridge (1998)
12. Kuipers, B.: The Spatial Semantic Hierarchy. Artificial Intelligence 119, 191–233 (2000)
13. Loncaric, S.: A survey of shape analysis techniques. Pattern Recognition 31(8), 983–1001 (1998)
14. Montemerlo, M., Roy, N., Thrun, S.: Perspectives on standardization in mobile robot programming. In: Proceedings of the IEEE/RSJ International Conference on Intelligent Robots and Systems (2003)
15. Moravec, H.P.: Sensor fusion in certainty grids for mobile robots. AI Magazine 9, 61–74 (1988)
16. Mozos, O.M., Stachniss, C., Burgard, W.: Supervised learning of places from range data using AdaBoost. In: Proceedings of the IEEE International Conference on Robotics and Automation, Barcelona, Spain, pp. 1742–1747 (2005)
17. Oore, S., Hinton, G.E., Dudek, G.: A mobile robot that learns its place. Neural Computation 9(3), 683–699 (1997)
18. O'Rourke, J.: Computational Geometry in C, 2nd edn. Cambridge University Press, Cambridge (1998)
19. Persson, M., Duckett, T., Lilienthal, A.J.: Virtual sensors for human concepts - building detection by an outdoor mobile robot. Robotics and Autonomous Systems 55(5), 383–390 (2007)
20. Pronobis, A., Caputo, B., Jensfelt, P., Christensen, H.I.: A discriminative approach to robust visual place recognition. In: Proceedings of the IEEE/RSJ International Conference on Intelligent Robots and Systems, Beijing, China (2006)
21. Pronobis, A., Mozos, O.M., Caputo, B.: SVM-based discriminative accumulation scheme for place recognition. In: Proceedings of the IEEE International Conference on Robotics and Automation, Pasadena, California, USA (2008)
22. Russ, J.C.: The Image Processing Handbook. CRC Press, Boca Raton (1992)
23. Sack, D., Burgard, W.: A comparison of methods for line extraction from range data. In: Proceedings of the IFAC Symposium on Intelligent Autonomous Vehicles, Lisabon, Portugal (2004)
24. Sousa, P., Araujo, R., Nunes, U.: Real-time labeling of places using support vector machines. In: Proceedings of the IEEE International symposium on industrial electronics, Vigo, Spain (June 2007)
25. Topp, E.A., Christensen, H.I.: Topological modelling for human augmented mapping. In: Proceedings of the IEEE/RSJ Conference on Intelligent Robots and Systems, Beijing, China (October 2006)
26. Torralba, A., Murphy, K.P., Freeman, W.T., Rubin, M.A.: Context-based vision system for place and object recognition. In: Proceedings of the International Conference on Computer Vision (2003)
27. Treptow, A., Masselli, A., Zell, A.: Real-time object tracking for soccer-robots without color information. In: Proceedings of the European Conference on Mobile Robots (2003)
28. Viola, P., Jones, M.J.: Robust real-time object detection. In: Proceedings of IEEE Workshop on Statistical and Theories of Computer Vision (2001)

Chapter 4
Topological Map Extraction with Semantic Information

4.1 Introduction

In the previous chapter we saw how a robot can classify its pose in an indoor environment into a semantic class. The different semantic classes represented typical divisions of the environment such as corridors, rooms or doorways. This chapter will show how a robot can extract a topological map from the environment using the previous semantic labeling.

Topological maps have been quite popular in the robotics community because they are believed to be cognitively more adequate, since they can be stored more compactly than geometric maps, and can also be communicated more easily to users of a mobile robot. Many researchers have considered the problem of building topological maps of the environment from the data gathered with a mobile robot. However, few techniques exist that permit semantic information to be added to these maps.

In this chapter, we consider the problem of learning topological maps with semantic information from occupancy grid maps that were obtained with a mobile robot in an indoor environment using range data. The approach is based on the assumption that indoor environments, like the one depicted in Fig. 4.1(a), can be typically decomposed into areas with different functionalities such as rooms, corridors and doorways, and that these areas build the vertices of a topological graph. The connections of the vertices are given by the neighborhood of the regions in the occupancy map. For example, a doorway is typically connected to two rooms, two corridors, or to a room and a corridor. Figure 4.1(b) depicts a possible topological representation for the map in Fig. 4.1(a)

Throughout this chapter we assume that the robot is given a map of the environment in the form of an occupancy grid. The main idea is to decide about the semantic label of each free cell using local and neighboring information. By local information we mean the set of geometrical features the robot obtains from a laser observation at a concrete location (cf. Chap. 3). By neighboring information we refer to the semantic information from the neighboring locations.

Ó.M. Mozos: Semantic Labeling of Places with Mobile Robots, STAR 61, pp. 35–56.
springerlink.com

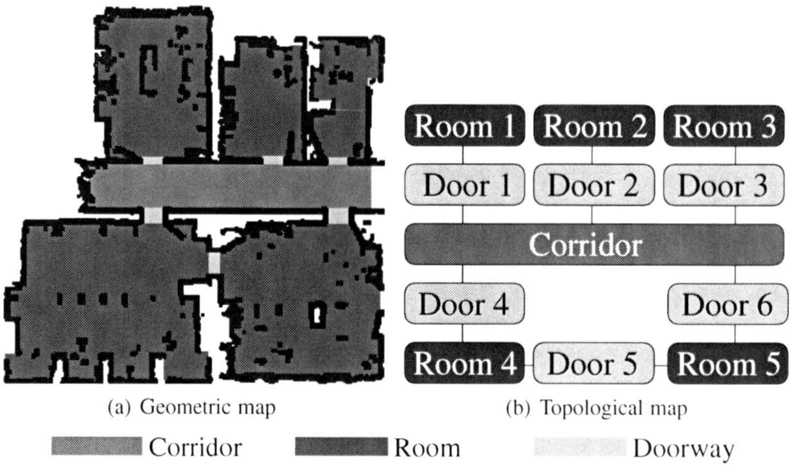

Fig. 4.1 (**a**) Geometric map of a typical indoor environment with rooms, doorways, and a corridor, depicted in colors/gray levels. (**b**) Corresponding semantic-topological map.

Two different methods are presented which use both local and neighboring information for the final classification. The first approach determines the semantic class of each unoccupied cell of a grid map. This is achieved by simulating a range scan of the robot given it is located in that particular cell, and then classifying this scan into one of the semantic classes. Examples for typical simulated range scans obtained in an office environment were shown in Chap. 3 (cf. Fig. 3.2). The classification is then done using a sequence of classifiers learned with the AdaBoost algorithm arranged in a probabilistic decision list in a similar way as introduced in Sect. 3.3. However, in this chapter we present some modifications in the learning and classification process which permit the use of probability values for the different classifications. Finally, to remove noise and clutter from the resulting classifications, we apply an approach denoted as probabilistic relaxation labeling. This method corrects the classification at each location taking into account the semantic class of neighboring positions.

The second method for the classification is based on associative Markov networks (AMNs). In this case, the semantic classification at one position inside the map is done using simultaneously the local information and the relation between semantic labels from neighboring positions. We apply a variant of AMNs called instance-based associative Markov networks (iAMNs). This concrete approach combines AMNs with nearest-neighbor techniques.

Experimental results shown in this chapter illustrate that these methods can determine the semantic-topological map of an environment with high recognition rates. We also present results that illustrate that this approach can even construct a topological map of an environment from which no training data was available. Finally, we extend the set of simple features used in Chap. 3 with new ones. As the experimental results illustrate, the newly created set provide better classification results.

The rest of the chapter is organized as follows. In the next section we introduce the application of the generalized AdaBoost for the concrete task of place recognition. A probabilistic version of a decision list is presented in Sect. 4.3. The extended set of geometric features is described in Sect. 4.4. The probabilistic relaxation approach is presented in Sect. 4.5. Instance-based associative Markov networks are introduced in Sect. 4.6. Section 4.7 describes the method used to extract semantic regions and to create the final topological map. In Sect. 4.8, experimental results are presented. We discuss related work in Sect. 4.9. Finally, we conclude in Sect. 4.10.

4.2 Generalized AdaBoost

In this chapter we use the variant of the AdaBoost algorithm known as generalized AdaBoost [19]. As introduced in Sect. 2.2.2, this version has several advantages over the original AdaBoost algorithm. In addition, its output can be easily converted into a confidence value.

The input to the generalized AdaBoost is also composed of a set of labeled training examples $(x_n, y_n), n = 1, \ldots, N$. However, the label for the examples is in this case $y_n = +1$ when x_n is positive, and $y_n = -1$ when x_n is negative. Similar to the original AdaBoost, during the different iterations $t = 1, \ldots, T$ the algorithm selects a weak classifier with small error in the weighted training examples. The weight distribution D_t is changed on each iteration to give more importance to the most difficult examples. The final strong classifier is composed of a weighted majority sum of the selected weak hypotheses.

Following the approach presented in Sect. 3.2, each weak classifier is based on single-valued features f_j and has the form

$$h_j(x) = \begin{cases} +1 & \text{if } p_j f_j(x) < p_j \theta_j \\ -1 & \text{otherwise} . \end{cases} \tag{4.1}$$

Equation (4.1) differs from (3.1) in the output for a negative classification, which in this case is -1. The final generalized AdaBoost algorithm modified for the concrete task of place labeling is given in Fig. 4.2.

Using the generalized version of the AdaBoost algorithm shown in Fig. 4.2, and following the method suggested in [5], we can additionally compute a confidence value $C^+ \in [0, 1]$ for a positive binary classification of a new example as

$$C^+ = P(y = +1 \mid x) = \frac{e^{F(x)}}{e^{-F(x)} + e^{F(x)}} , \tag{4.2}$$

where $F(x)$ is the output of the algorithm according to Fig. 4.2. If the example is classified as negative, the positive confidence value can be calculated as

- Input:
  - Set of N labeled examples $(x_1, y_1), \ldots, (x_N, y_N)$ with $y_n = +1$ if the example x_n is positive, and $y_n = -1$ if the example x_n is negative
  - Integer T specifying the number of iterations
- Initialize weights $D_1(n) = \frac{1}{2l}$ for positive examples, and $D_1(n) = \frac{1}{2m}$ for negative examples, where l is the number of positive examples and m the number of negative ones.
- For $t = 1, \ldots, T$
  1. Normalize the weights $D_t(n)$
  $$D_t(n) = \frac{D_t(n)}{\sum_{i=1}^{N} D_t(i)} .$$
  2. For each feature f_j train a weak classifier h_j using the distribution D_t.
  3. For each classifier h_j calculate
  $$r_j = \sum_{n=1}^{N} D_t(n) y_n h_j(x_n) ,$$
  where $h_j(x_n) \in [-1, +1]$.
  4. Choose the classifier h_j that maximizes $|r_j|$ and set $(h_t, r_t) = (h_j, r_j)$.
  5. Update the weights
  $$D_{t+1}(n) = D_t(n) \exp(-\alpha_t y_n h_t(x_n)) ,$$
  where $\alpha_t = \frac{1}{2} \log(\frac{1+r_t}{1-r_t})$.
- The final strong classifier is given by
$$H(x) = \operatorname{sign}(F(x)) ,$$
where
$$F(x) = \sum_{t=1}^{T} \alpha_t h_t(x) .$$

Fig. 4.2 The generalized version of the AdaBoost algorithm for place labeling using laser-based features.

$$C^+ = P(y = +1 \mid x) = 1 - C^- , \tag{4.3}$$

with

$$C^- = P(y = -1 \mid x) = \frac{e^{-F(x)}}{e^{-F(x)} + e^{F(x)}} . \tag{4.4}$$

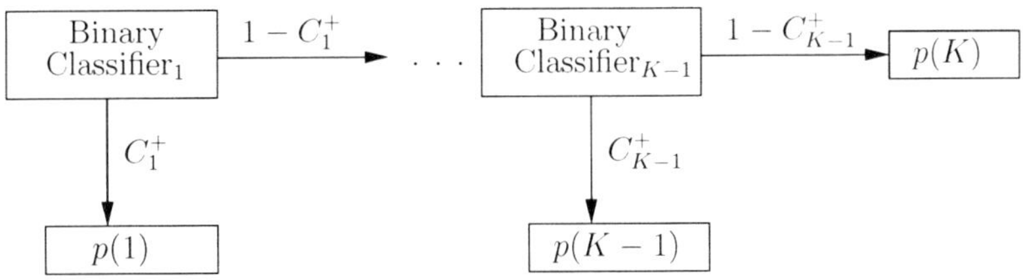

Fig. 4.3 A decision list classifier for K classes using binary classifiers. The output of each binary classifier $p(k)$ contains the probability that the classified example belongs to the k-th class.

4.3 Probabilistic Decision List

Extending the ideas introduced in Sect. 3.3, we use a probabilistic decision list to create a classifier for multiple classes. Each element of such a list represents one binary classifier which determines if an example belongs to one specific class. In addition, each binary classifier outputs a confidence value C_k^+ for a positive classification of its class k. Figure 4.3 illustrates the structure of a probabilistic decision list.

In this decision list, each test example is fed into the first binary classifier, which outputs a confidence value C^+ for a positive classification. The example is also passed to the next binary classifier, but with a negative confidence value $1 - C^+$. This process is repeated until the last element in the list. The complete output of the decision list is represented by a histogram P. In this histogram, the bin $p(k)$ stores the probability that the classified location belongs to the k-th class according to the sequence of classifiers in the decision list. Let C_k^+ refer to the positive confidence value of the k-th binary classifier in our decision list. Then, the probability that the example to be classified belongs to the k-th class is given by the bin $p(k)$ of the histogram P computed as

$$p(k) = C_k^+ \prod_{j=1}^{k-1} (1 - C_j^+), \tag{4.5}$$

whereas for the confidence value C_K^+ of the last bin holds $C_K^+ = 1$ according to the structure of the decision list in Fig. 4.3. An example of a histogram for six classes is illustrated in Fig. 4.4.

To select the order of the different binary classifiers we try all possible combinations and choose the one with best classification rates. Each binary classifier is trained in a one-against-all fashion, selecting one class as positive examples and the rest of the classes as negative examples. This is similar to the approach introduced in Sect. 3.3.

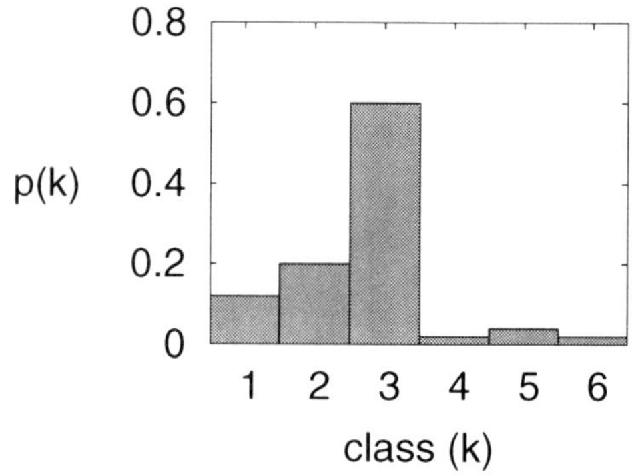

Fig. 4.4 An example of a classification output for the decision list of Fig. 4.3 using six classes.

4.4 New Geometrical Features from Sensor Range Data

As explained in Chap. 3, in order to classify each free cell in the occupancy grid map we simulate a range scan in its position ray-casting in the map. The simulated scans correspond to a robot equipped with a 360° degree field of view laser sensor. Each simulated laser observation then consists of 360 beams. Each training example for the AdaBoost algorithm consists of the features extracted from the laser observation, together with its classification. Moreover, we assume that the classification of the training examples is given in advance. The single-valued features used in the AdaBoost algorithm are geometrical features used for shape analysis [7, 9, 13, 15, 18]. The features are selected to be rotationally invariant to make the classification of a pose dependent only on the (x,y)-position of the robot and not on its orientation. A feature f is defined as a function that takes as argument one observation $z \in Z$ and returns a real value: $f(Z) \rightarrow \mathbb{R}$, with Z being the set of all observations. In this chapter we apply an extended set of the features introduced in Sect. 3.4. The following list extends the original set A, which contains features calculated from the raw beams of z:

11. Average of the relation between the length of two consecutive beams.
12. Standard deviation of the relation between the length of two consecutive beams.
13. Average of normalized beam length.
14. Standard deviation of normalized beam length.
15. Number of relative gaps.
16. Kurtosis.

The set B, which corresponds to the geometrical features extracted from the polygonal approximation Pol(z) of the observation z, is also extended with the following new features:

14. The circularity of Pol(z).
15. The normalized circularity of Pol(z).

16. The average normalized distance between the centroid and the shape boundary of $\mathrm{Pol}(z)$.
17. The standard deviation of the normalized distance between the centroid and the shape boundary of $\mathrm{Pol}(z)$.

In the experimental section we will see that these additional features improve the robustness of the resulting classifier. The complete lists of features, together with their mathematical definition can be found in Appx. A and Appx. B.

4.5 Probabilistic Relaxation Labeling

The first approach that we use in this chapter to extract topological maps determines the semantic class of each unoccupied cell of the grid. This is achieved by simulating a range scan of the robot given it is located at that particular cell, and then labeling this scan into one of the semantic classes using a probabilistic decision list as presented in Sect. 4.3. This process results in an occupancy map with a semantic label for each free cell. However, the final maps usually contain some errors in the classification. To smooth the final classification of each cell, we apply the probabilistic relaxation labeling method introduced in [16]. This method changes (or maintains) the label of a cell according to the labels of its neighborhood.

The probabilistic relaxation labeling problem is defined as follows. Let $\mathrm{G} = (\mathrm{V},\mathrm{E})$ be a graph consisting of nodes $\mathrm{V} = \{v_1,\ldots,v_N\}$ and edges $\mathrm{E} \subseteq \mathrm{V} \times \mathrm{V}$. Let furthermore $\mathrm{Y} = \{y_1,\ldots,y_K\}$ be a set of labels. We assume that every node v_i stores a probability distribution about its label. This distribution is represented by a histogram P_i. Each bin $p_i(k)$ of that histogram stores the probability that the node v_i has the label k. Thus, $\sum_{k=1}^{K} p_i(k) = 1$. For each node v_i, $\mathrm{Ne}(v_i) \subset \mathrm{V}$ denotes its neighborhood, which consists of the nodes $v_j \neq v_i$ that are connected to v_i. Each neighborhood relation is represented by two values. Whereas the first one describes the compatibility between the labels of two nodes, the second one represents the influence between the two nodes. The term $\mathrm{R} = \{r_{ij}(k,k') \mid v_j \in \mathrm{Ne}(v_i)\}$ defines the compatibility coefficients between the label k of node v_i and the label k' of v_j. Additionally, we define $\mathrm{O} = \{o_{ij} \mid v_j \in \mathrm{Ne}(v_i)\}$ as the set of weights indicating the influence of node v_j on node v_i.

Given an initial estimation for the probability distribution over labels $P_i^{(0)}$ for the node v_i, the probabilistic relaxation method iteratively computes estimates $P_i^{(r)}$, $r = 1,2,\ldots$, based on the initial probabilities $p_i^{(0)}(k)$, the compatibility coefficients R, and the weights O, in the form

$$p_i^{(r+1)}(k) = \frac{p_i^{(r)}(k)\left[1+q_i^{(r)}(k)\right]}{\sum_{k'=1}^{K} p_i^{(r)}(k')\left[1+q_i^{(r)}(k')\right]}, \tag{4.6}$$

where

$$q_i^{(r)}(k) = \sum_{j=1}^{N} o_{ij} \left[\sum_{k'=1}^{K} r_{ij}(k,k') p_j^{(r)}(k') \right] . \tag{4.7}$$

Note that the compatibility coefficients $r_{ij}(k,k') \in [-1,1]$ do not need to be symmetric. A value $r_{ij}(k,k')$ close to -1 indicates that label k' is unlikely at node v_j when label k occurs at node v_i, whereas values close to $+1$ indicate the opposite. A value of exactly -1 indicates that the relation is not possible, and a value of exactly $+1$ means that the relation always occurs.

Probabilistic relaxation provides a framework for smoothing but does not specify how the compatibility coefficients are computed. In this work, we apply the coefficients as defined in [26] as

$$r_{ij}(k,k') = \begin{cases} \frac{1}{1-p_i(k)} \left(1 - \frac{p_i(k)}{p_{ij}(k|k')}\right) & \text{if } p_i(k) < p_{ij}(k \mid k') \\ \frac{p_{ij}(k|k')}{p_i(k)} - 1 & \text{otherwise} , \end{cases} \tag{4.8}$$

where $p_{ij}(k \mid k')$ is the conditional probability that node v_i has label k given that node $v_j \in \mathrm{Ne}(v_i)$ has label k'. Each of the values $p_i(k)$ and $p_{ij}(k \mid k')$ are pre-calculated only once and remain the same during the iterations of the relaxation process. The coefficients R remain the same as well.

Now we describe how to apply this method for the spatial smoothing of the classifications obtained by our classifier. To learn a topological map, we assume a given two-dimensional occupancy grid map in which each cell $m_{(x,y)}$ stores the probability that the cell is occupied. We furthermore consider the 8-connected graph induced by such a grid. Let $v_i = v_{(x,y)}$ be a node corresponding to a cell $m_{(x,y)}$ from the map. We then define a neighborhood $\mathrm{Ne}(v_{(x,y)})$ using the 8-connected cells to $v_{(x,y)}$ as described in [7].

For the initial probabilities $p_{(x,y)}^{(0)}(k)$, we use the output P of the probabilistic decision list as described in Sect. 4.3. This output is represented by a histogram in which each bin $p(k)$ indicates the probability that the pose belongs to class k. Furthermore, our set of labels Y is composed by four labels

$$\mathrm{Y} = \{\text{corridor}, \text{room}, \text{doorway}, \text{wall}\} .$$

For each node $v_{(x,y)}$ in the free space of the occupancy grid map, we calculate the expected laser scan by ray-casting in the map. We then classify the observation and obtain a probability distribution z over all the possible places according to (4.5). The classification output P for each pose (x,y) is used to initialize the probability distribution $P_{(x,y)}^{(0)}$ of node $v_{(x,y)}$. For the nodes lying in the free space, the probability $p_{(x,y)}^{(0)}(wall)$ of being a wall is initialized with 0. Accordingly, the nodes corresponding to occupied cells in the map are initialized with $p_{(x,y)}^{(0)}(wall) = 1$.

Each of the weights $o_{ij} \in \mathrm{O}$ is initialized with the value $\frac{1}{8}$, indicating that the eight neighbors v_j of node v_i are equally important. The compatibility coefficients are calculated using (4.8). The values $p_i(k)$ and $p_{ij}(k \mid k')$ are obtained from statistics in the given occupancy grid map corresponding to previously labeled training data.

4.6 Instance-Based Associative Markov Networks[1]

The second approach for topological map extraction presented in this chapter is based on associative Markov networks (AMNs). In particular, we use the instance-based associative Markov networks (iAMNs) introduced in [25]. The idea behind iAMNs is to combine the advantage of instance-based nearest-neighbor (NN) classification with the AMN approach to obtain a collective classifier that is not restricted to the linear separability requirement.

4.6.1 Associative Markov Networks

An associative Markov network is an undirected graphical model in which no assumption is made about the direction of the causality between nodes in the graph. We restrict ourselves to the case of discrete variables, that is, each variable $y_i \in \mathrm{Y}$ corresponds to a set of K possible labels $y_i \in \{1, \ldots, K\}$. Thus, we define a Markov random field as an undirected graph $\mathrm{G} = (\mathrm{V}, \mathrm{E})$ where the set of nodes V represents discrete variables, and the edges E refer to the relations between them [22]. An AMN can be divided into a subset of cliques Q, where each clique $q \in \mathrm{Q}$ is associated with a subset $\mathrm{Y}_q \in \mathrm{Y}$. The nodes in a clique Y_q form a fully connected subgraph.

Each clique q is accompanied by a potential $\phi_c(y_q)$ which associates a non-negative value to the variable assignment y_q. We work with pairwise associative Markov networks [22], where all of the cliques involved are either a single node, or a pair of nodes (1-clique or 2-clique). In a pairwise AMN with edges $\mathrm{E} = \{(ij) \mid i < j\}$, the nodes and edges are associated with potentials $\phi_i(y_i)$ and $\phi_{ij}(y_i, y_j)$ respectively.

In an AMN, each node y_i can be assigned a feature vector $x_l \subset \mathbb{R}^L$, which describes the properties of the object represented by that node. Similarly, a feature vector $x_{ij} \in \mathbb{R}^{L'}$ can be assigned to each edge $(ij) \in \mathrm{E}$. The feature vector x_{ij} indicates the properties that describe the relation between the objects represented by the nodes y_i and y_j. The node potentials are functions of the node feature vectors x_i, similarly the edge potentials are functions of the edge feature vectors x_{ij}. The resulting network defines the distribution

$$\log P_w(y|x) = \sum_{i=1}^{N} \sum_{k=1}^{K} (w_n^k \cdot x_i) y_i^k + \sum_{e=(ij)\in \mathrm{E}} \sum_{k,k'=1}^{K} (w_e^{k,k'} \cdot x_{ij}) y_i^k y_j^{k'} - \log Z_w(x) \,, \quad (4.9)$$

[1] The work presented in this section originated from a collaboration with Rudolph Triebel.

where N is the total number of nodes in the graph, and $Z_w(x)$ is a partition function that depends on the parameters w and features x, but not on the labels y.

The main task in an associative Markov network consists of finding the assignment $y \in \mathrm{Y}$ that maximizes $\log_w P(y|x)$. This is actually a maximum a posteriori (MAP) assignment that can be formulated as a linear program [22].

4.6.2 *Feature Vector Transformation*

The main drawback of the AMN classifier, which is based on the log-linear model, is that it separates the classes linearly. This assumes that the features are separable by hyper-planes, which is not justified in all applications. This restriction does not hold for instance-based classifiers such as the nearest-neighbor, in which a query data point $\tilde{p}$ is assigned to the label that corresponds to the training data point p whose features x are closest to the features $\tilde{x}$ of $\tilde{p}$. In the learning step, the NN classifier simply stores the training data set and does not compute a reduced set of training parameters.

To combine the advantage of instance-based NN classification with the AMN approach, we convert the feature vector $\tilde{x}$ of length L pertaining to query point $\tilde{p}$ using the transform $\tau : \mathbb{R}^L \rightarrow \mathbb{R}^K$ given by

$$\tau(\tilde{x}) = (t_1 = d(\tilde{x}, \hat{x}_1), \ldots, t_K = d(\tilde{x}, \hat{x}_K)) , \qquad (4.10)$$

where K is the number of classes, and $\hat{x}_k$ denotes the training example with label k closest to $\tilde{x}$. In addition, the function $d(\cdot,\cdot)$ calculates the distance in feature space. Using this transformation the resulting features are more easily separable by hyper-planes. An example is given in Fig. 4.5. Here, the top image depicts the training and test data for a two class problem, in which the length of the feature vector $x = (x_1, x_2)$ is two. The classification of the test data (triangles) is shown as lines connecting each training example with the closest example in the ground truth (squares). This nearest neighbor classification results in very few errors. However, it seems difficult to separate the test data into the two classes to which they pertain using a hyperplane (a line in this case). The bottom image of Fig. 4.5 shows the training examples in the transformed space using the transformation given by $\tau(\tilde{x}) = (t_1, t_2)$, with $t_1 = d(\tilde{x}, \hat{x}_1)$, and $t_2 = d(\tilde{x}, \hat{x}_2)$. The linear separability is improved in the transformed space.

Additionally, the M nearest neighbors can be used in the transform function. For this, we compute the M nearest distances to each of the classes $k = 1, \ldots, K$. The final transformation $\tau_M : \mathbb{R}^L \rightarrow \mathbb{R}^{K \cdot M}$ is given by

$$\tau_M(\tilde{x}) = (d(\tilde{x}, \hat{x}_1^1), \ldots, d(\tilde{x}, \hat{x}_1^M), \ldots, d(\tilde{x}, \hat{x}_K^1), \ldots, d(\tilde{x}, \hat{x}_K^M)) . \qquad (4.11)$$

The resulting model, first introduced in [25], is called instance-based associative Markov network (iAMN).

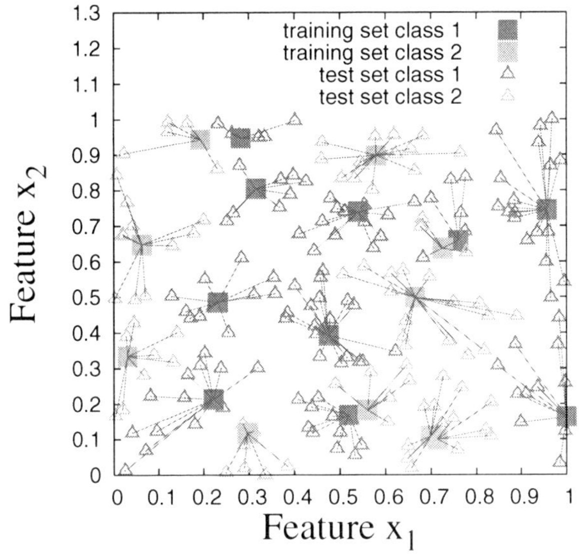

(a) original feature space

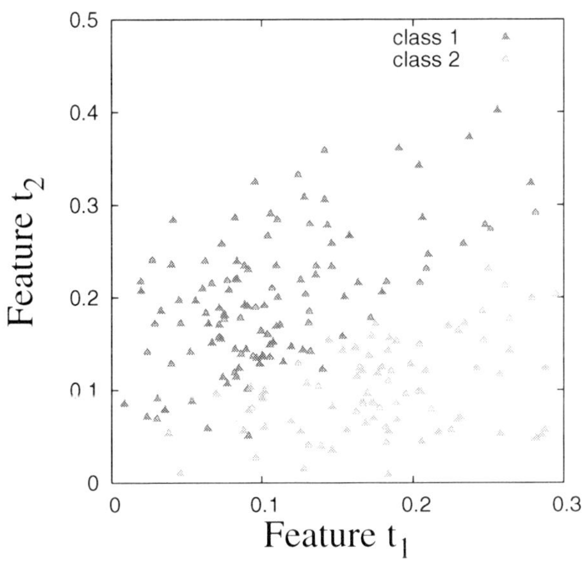

(b) original feature space

Fig. 4.5 Example of the feature transform τ for a two-class problem with two features. (**a**) Training and test data in the original feature space. The classification of the test data (*triangles*) is shown as lines connecting each training example with the closest example in the ground truth (*squares*). (**b**) The transformation τ is applied to the test data (*triangles*). The transformed examples are more easily separable.

4.6.3 Feature Selection

One of the problems when classifying points represented by range data consists of selecting the size of the feature vectors. As we showed in the experiments of Chapter 3, the number of possible features that can be used to represent each data point is usually very large and can easily be in the order of hundreds. This problem is known as the *curse of dimensionality*. There are at least two reasons to try to reduce the size of the feature vector. The most obvious one is the computational complexity, which in our case, is also the most critical, since we have to learn and infer in networks with thousands of nodes. Another reason is that although some features may carry a good classification when treated separately, maybe there is a little gain when combined together if they are very correlated [23]. The goal is thus to reduce the size of the feature vectors when used with the iAMN and, at the same time, to try to maintain their class discriminatory information.

The reduction on the numbers of features used for the classification of places is somehow implicit in the AdaBoost classifiers, since the final number of weak classifiers T can be determined, and each selected weak classifier represents a feature (cf. Sect 4.2). The problem is that the same feature can appear multiple times with different thresholds and different priorities, which makes it difficult to decide which are the best original features.

In this section we follow an alternative approach for selecting features. We apply a scalar feature selection procedure which uses a class separability criterion and incorporates correlation information. The selection is independent of the classification algorithm that will use the features (iAMN in our case). These kinds of methods are also denoted as *filters*. A filter relies on general characteristics of the data to evaluate and select feature subsets without involving any classification algorithm [8].

As separability criterion S, we use the Fisher's discrimination ratio (FDR) extended to the multi-class case [23]. For a scalar feature f and K classes $\{y_1, \ldots, y_K\}$, $S(f)$ can be defined as

$$S(f) = FDR_f = \sum_{i=1}^{K} \sum_{j \neq i}^{K} \frac{(\mu_i - \mu_j)^2}{\sigma_i + \sigma_j}, \tag{4.12}$$

where the μ_i and σ_i refer respectively to the mean and variance of the class i. Additionally, the cross-correlation coefficient between any two features f and f' given N training examples is defined as

$$\rho_{ff'} = \frac{\sum_{t=1}^{N} f_n f'_n}{\sqrt{\sum_{n=1}^{N} f_n^2 \sum_{n=1}^{N} f'^2_n}}, \tag{4.13}$$

where f_n denotes the value of the feature f in the training example n. Finally, the selection of the best $L^* \subset L$ features involves the steps shown in Fig. 4.6.

After the scalar feature selection, the learning and inference steps on the instance-based associative Markov network are carried out. Further details on the inference process can be found in [25].

- Select the first feature f_1 as

$$f_1 = \operatorname*{argmax}_{f} S(f).$$

- Select the second feature f_2 as

$$f_2 = \operatorname*{argmax}_{f \neq f_1} \left\{ w_1 S(f) - w_2 |\rho_{f_1 f}| \right\},$$

where w_1 and w_2 are weighting factors.

- Select f_{l^*}, $l^* = 3, \ldots, L^*$, such that

$$f_{l^*} = \operatorname*{argmax}_{f \neq f_r} \left\{ w_1 S(f) - \frac{w_2}{l^* - 1} \sum_{r=1}^{l^*} |\rho_{f_r f}| \right\}, \quad r = 1, 2, \ldots, l^* - 1$$

Fig. 4.6 Feature selection algorithm according to [23].

4.7 Region Extraction and Topological Mapping

After applying any of the previous two approaches, we obtain an occupancy grid map in which each free cell contains a distribution over the set of its possible labels. From this map we extract complete regions that correspond to places in the environment.

We define a region on a adjacency graph G as a set of 8-connected nodes with the same class y. For each label $y \in \{\text{corridor}, \text{room}, \text{doorway}\}$, regions are extracted from the adjacency graph using an algorithm for extracting connected regions [17]. Each extracted region is assigned a different identifier. The connections between regions are extracted using a similar algorithm [7]. Finally, a new topological graph $\mathrm{G} = (\mathrm{V}, \mathrm{E})$ is constructed in which each node $v_i \in \mathrm{V}$ represents a region and each edge $e_i \in \mathrm{E}$ represents a connection. Additionally, we add to each node v_i information about the properties of the region which represents: area, centroid, and major and minor axes of the ellipse approximation of the region. The major and minor axes are vectors which represent the elongation of the region and its orientation. The topological graph together with the region properties form the final topological map. We finally apply a heuristic region correction to the topological map to increase the classification rate as follows:

1. We mark each region corresponding to a room or a corridor whose size does not exceed a given threshold of 1 m^2 compared to the training set as a classification error, and assign it the label of one of its connected regions.
2. We mark each region previously labeled as doorway, and whose size does not exceed a given threshold of 0.1 m^2 or that is connected to only one region as a false classification. Then we assign these marked regions the label of one of their connected regions.

The different thresholds used in the heuristics are obtained from statistics in the training set.

4.8 Experimental Results

The approaches described above have been implemented and tested using occupancy maps obtained from real environments. The laser range data used for the training and classification steps were simulated using the Carnegie Mellon Robot Navigation Toolkit (CARMEN) [2, 14]. The goal of the experiments is to demonstrate that we can construct a semantic-topological map of typical indoor environments using only laser range data. We first apply our method using probabilistic relaxation. Additionally, we analyze whether this method can be used to create a topological map of an environment for which no training data were available. Furthermore, we analyze the improvement of the AdaBoost-based decision list classifier using the new set of features. Finally we present one experiment in which iAMNs are used to train and classify an indoor environment.

4.8.1 Results Using Relaxation Labeling

The first experiment was performed using data obtained in the office environment of building 79 at the University of Freiburg. This environment contains rooms, doorways and a corridor, which has a length of approximately 22 meters. For the sake of clarity we give the result of the obtained classification by separating the environment into two parts. The left half of the environment contains the poses used as training examples, and the right half of the environment was used for test classification and for the topological map creation as shown in Fig 4.7(a).

We first applied a probabilistic decision list (cf. Sec. 4.3) to classify the free cells in the occupancy grid map. The list was formed by the best combination of binary classifiers, which in this case was corridor-room. This sequence correctly classified 97.27% of the test examples. The classification is depicted as colors/gray levels in Fig. 4.7(b).

After the sequential classification, the probabilistic relaxation method explained in Sect. 4.5 was applied for 50 iterations. This method generates more compact regions and eliminates noise. The result is illustrated in the Fig. 4.7(c). Finally, the topological map is created using the connections between regions. As can be seen in Fig. 4.7(c), some regions detected as doorways (marked with circles) do not correspond to real doorways. After applying the heuristics described in Sect. 4.7 on the corresponding topological map, these false doorways are eliminated. Furthermore, the two left rooms situated above the corridor are detected as only one region. That is due to the fact that the doorway in between was not completely detected. Thus, the two rooms remain connected and are classified as only one region. The final topological map, depicted in Fig. 4.7(d), has a final classification rate of 98.95% in the free cells.

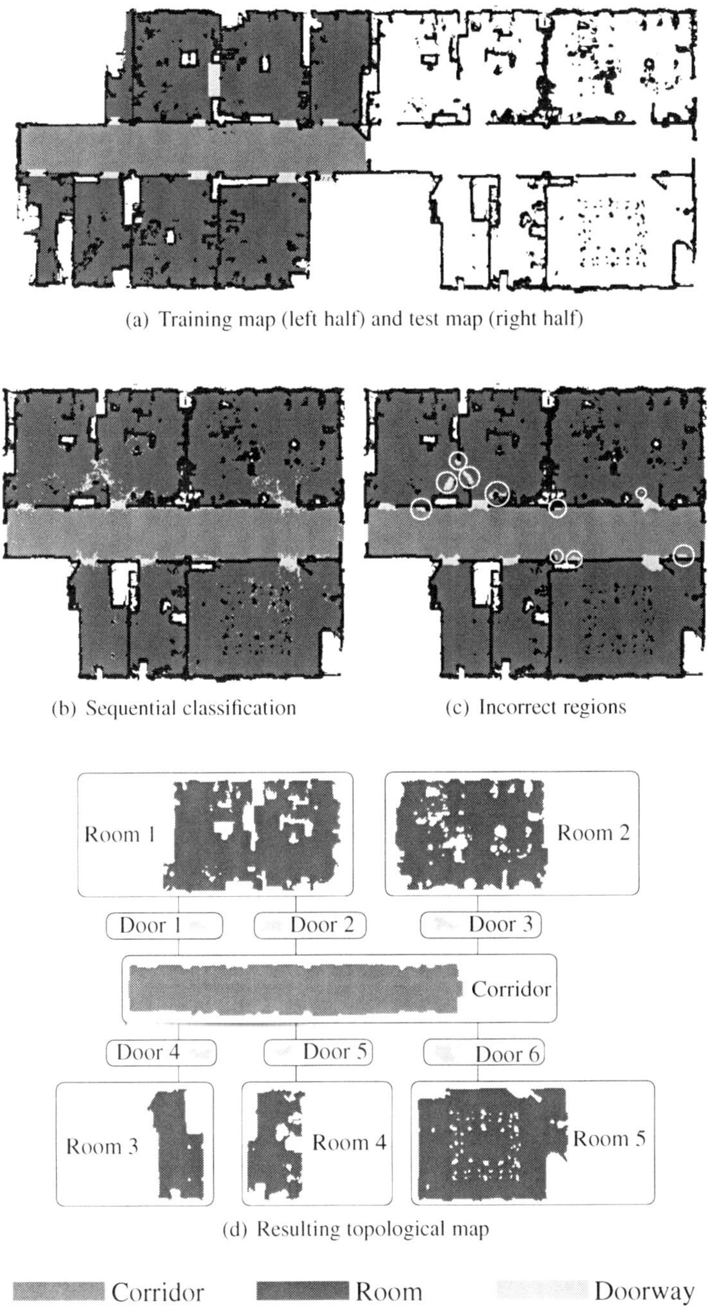

(a) Training map (left half) and test map (right half)

(b) Sequential classification

(c) Incorrect regions

(d) Resulting topological map

Fig. 4.7 (**a**) Training and test map of building 79 at the University of Freiburg. (**b**) Result of applying the decision list with a classification rate of 97.27%. (**c**) Result of applying relaxation and the detection of incorrect labeled regions (*white circles*). (**d**) Final topological map with the corresponding regions.

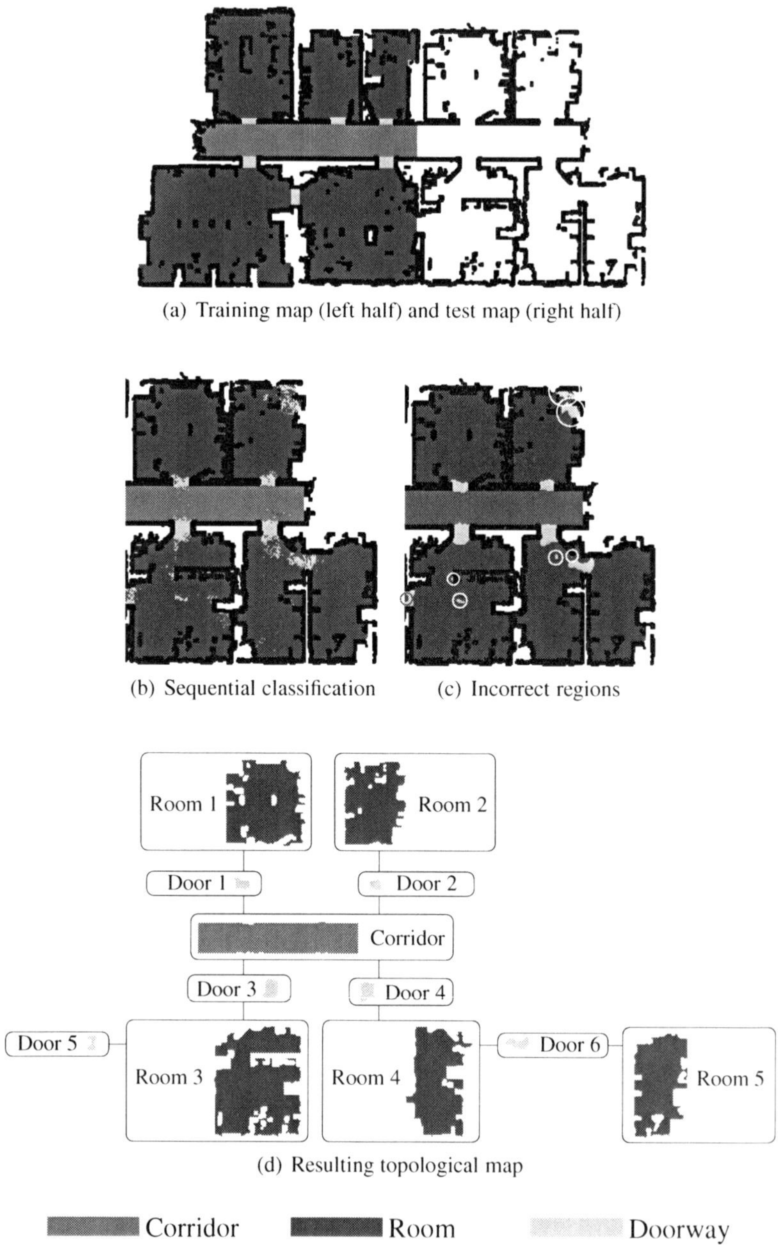

(a) Training map (left half) and test map (right half)

(b) Sequential classification (c) Incorrect regions

(d) Resulting topological map

Fig. 4.8 (**a**) Training and test map of the building 52 at the University of Freiburg. (**b**) Result of applying the decision list with a classification rate of 97%. (**c**) Result of applying relaxation and the detection of incorrect labeled regions (*white circles*). (**d**) Final topological map with the corresponding regions.

In a second experiment we created a topological map of the right part of the office environment of building 52 at the University of Freiburg. The complete occupancy grid map of this environment is shown in Fig. 4.8(a). The length of the corridor in this environment is approximately 20 meters. After applying the decision list classifier room-corridor, the classification of the test set was 97%. As in the previous experiment, we applied the relaxation process for 50 iterations as well as the heuristics for region correction of Sect. 4.7. The final result gives a classification rate of 98.66% in the free cells. The different steps of the process are illustrated as colors/gray levels in Fig. 4.8. Opposite to the previous experiment, the doorway between the two right-most rooms under the corridor is correctly detected, as can be shown in Fig. 4.8(c). Therefore, the rooms are labeled as two different regions in the final topological map as shown in Fig. 4.8(d).

4.8.2 Application to New Indoor Environments

This experiment is designed to analyze whether our approach based on boosting and relaxation labeling can be used to create a topological map of a new environment from which no training data were available. To carry out the experiment we trained a decision list classifier using the training examples of the maps shown in Fig. 4.7(a) and Fig. 4.8(a) at different scales. In this way, we obtained a classifier with a better generalization. The resulting classifier was then evaluated on scans simulated in the map denoted as *SDR site B* in the Radish repository [10]. This map represents an empty building in Virginia, USA. The corridor is approximately 26 meters long. The whole process for obtaining the topological map is depicted in Fig. 4.9. We use the sequence corridor-doorway which gives a first classification of 92.36%. As can be seen in Fig. 4.9(c), rooms number 11 and 30 are originally part of the corridor, and thus falsely classified. Moreover, the corridor is detected as only one region, although humans potentially would prefer to separate it into six different corridors: four horizontal and two vertical. Doorways are very difficult to detect with the sequential classifier. The majority of poses detected as doorways disappear after the relaxation process because they are very sparse. The main reason for the problem of doorway detection is that the training maps have different sizes and resolutions, and not all features are scale invariant. In the final topological map, 96.94% of the data points are correctly classified.

4.8.3 Results with Instance-Based Associative Markov Networks

In this experiment we applied our classification approach using iAMNs to the indoor environment corresponding to building 79 at the University of Freiburg. For efficiency reasons we used a grid resolution of 20 cm, which lead us to a graph containing 8088 nodes. Smaller resolutions result in much bigger networks, which are difficult to treat. As in the first experiment, the map was divided into two parts, the left one used for learning, and the right one used for classification purposes as

(a) SDR site B map

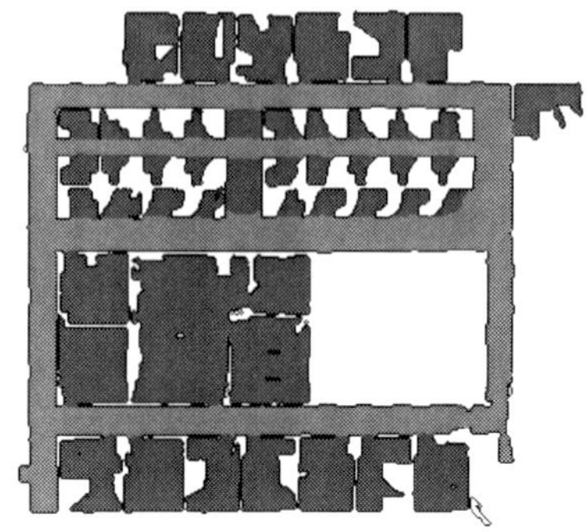

(b) Map after relaxation and region correction

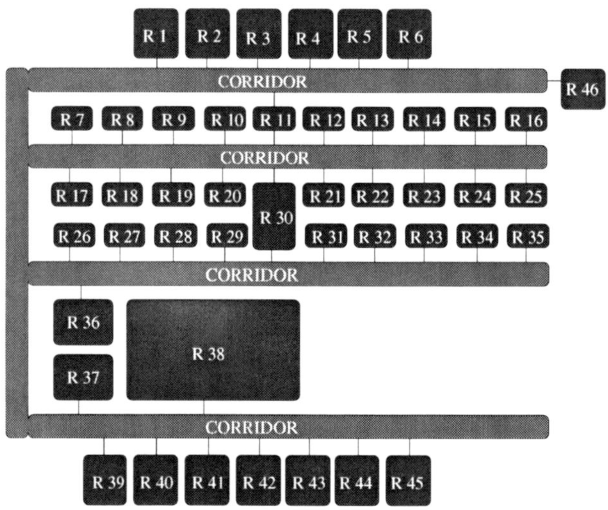

(c) Final topological map

Fig. 4.9 (**a**) Original map of the building. (**b**) Resulting classification after the relaxation an region correction. (**c**) Final topological map with semantic information. The regions are omitted in each node. The rooms are numbered left to right and top to bottom with respect the map in (a). For the sake of clarity, the corridor-node is drawn maintaining part of its original structure.

shown in Fig. 4.10. For each cell we calculate 203 geometrical features. This number was reduced to 30 applying the feature selection of Sect. 4.6.3. Figure 4.10(b) shows the resulting classification with a success rate of 97.6%, which is similar to the classification obtained using relaxation labeling. We can also see in the results that some doorways are lost in the final classification. The reason for this could be the low resolution of the map (20 cm) in comparison with the original

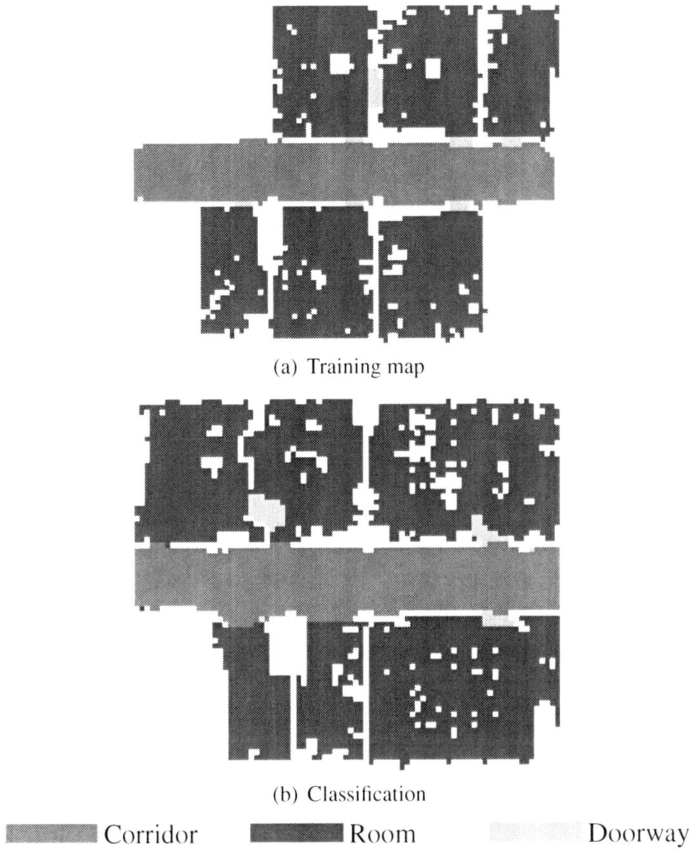

Fig. 4.10 (**a**) Training map of building 79 at the University of Freiburg. (**b**) Resulting classification using an iAMN with 30 selected features.

resolution (5 cm). However, maintaining the original resolution would lead us to a huge Markov network almost impractical to use.

4.8.4 Comparison of Feature Sets

In this final experiment, we compare the new extended feature set described in Sect. 4.4 with the one proposed Sect. 3.4. For this purpose, we trained an AdaBoost-based decision list for each of the feature sets using the training set shown in Fig. 4.7(a). The different sequential classifiers were then applied to the test set depicted in Fig. 4.7(b). The obtained classification results are shown in Table 4.1. As can be seen, the new extended feature-set provides better results in all of the experiments. This result indicates that the feature set has a major influence in the final classification. However, the advantage of AdaBoost is that we can keep adding

Table 4.1 Classification results of the new improved feature set compared to the one in Section 3.4.

Decision List	Feature set of Sect. 4.4 [%]	Feature set of Sect. 3.4 [%]
corridor-room	97.27	93.16
room-corridor	97.26	93.31
room-door	96.94	93.94
corridor-door	87.73	80.10
door-corridor	87.21	80.10
door-room	86.60	80.49

features to the set without worrying about the curse of dimensionality, since the boosting process will select only the necessary features.

4.9 Related Work

Different algorithms for extracting topological maps in indoor environments have been proposed in the past. For example, Kuipers and Byun [12] extract distinctive points in the map, which are defined as the local maximum of some measure. These points are used as nodes in a topological map. In their work, Kortenkamp and Weymouth [11] fuse vision and ultrasound information to determine topologically relevant places. Additionally, Shatkey and Kaelbling [20] apply a based learning approach based on hidden Markov models to learn topological maps in which the nodes represent points in the plane. Critical points are also found by Thrun [24], in this case using Voronoi diagrams. These critical points minimize the clearance locally, and are then used as nodes in a topological map. Also Beeson et al. [1] detect topological places with an extension of the Voronoi graph. Furthermore, Choset [3] encodes metric and topological information in a generalized Voronoi graph to solve the SLAM problem. The distinctive places extracted with the previous methods do not represent concrete spaces such as rooms or corridors although they can have some relation to them. In comparison, the technique described in this chapter applies a supervised learning method to identify complete regions in the map like corridors, rooms or doorways that have a direct relation with a human understanding of the environment.

Mathematical morphology is used in the work by Fabrizi and Saffiotti [4]. This method uses a disc as structuring element for the dilation and erosion operations. This approach extracts large open spaces from the map, but is quite sensitive to irregularities in the map.

Other works use vision sensors to distinguish places in an indoor environment. Tapus and Siegwart [21] use fingerprints extracted from images to create topological maps. In their work, Zivkovic et al. [27] create a higher level conceptual map with visual landmarks and geometric constraints. These approaches used features extracted from images that are quite specific to the environment in which the robot is located, which makes it difficult to generalize with new environments. In contrast to these works, the methods presented in this chapter have better generalization, since they used the geometrical properties of the different places.

In a recent work, Friedman et al. [6] use Voronoi Random Fields for extracting the topologies of occupancy grid maps. This work also uses simple features that are selected using boosting as characteristics for the nodes in a Markov random field. The approach is similar to the one in Sect. 4.6. However, in [6], only the points lying in the Voronoi diagram are used in the MRF, whereas we used all free locations in the map.

For related work about semantic place classification we refer the reader to Sect. 3.7.

4.10 Conclusion

In this chapter, we presented several approaches to create topological maps from indoor environments. The first one uses AdaBoost to learn a strong classifier for categorizing places into semantic classes such as rooms, doorways, and corridors. A probabilistic relaxation process is applied to the resulting classifications to reduce classification errors. The second approach is based on iAMNs together with scalar feature selection. After applying any of the previous methods the different regions and their connections are extracted. Each region corresponds to a place in the map such a corridor, room, or doorway.

Both methods has been implemented and evaluated on various maps from real-world environments. Experiments demonstrate that they are well-suited to creating topological maps from indoor environments, even without training the classifier for each environment.

References

1. Beeson, P., Jong, N.K., Kuipers, B.: Towards autonomous topological place detection using the extended voronoi graph. In: Proceedings of the IEEE International Conference on Robotics and Automation (2005)
2. CARMEN. Carnegie Mellon Robot Navigation Toolkit, `http://carmen.sourceforge.net/`
3. Choset, H.: Topological simultaneous localization and mapping (SLAM): Toward exact localization without explicit localization. IEEE Transactions on Robotics and Automation (2001)
4. Fabrizi, E., Saffiotti, A.: Extracting topology-based maps from gridmaps. In: Proceedings of the IEEE International Conference on Robotics and Automation, San Francisco, CA, pp. 2972–2978 (2000)
5. Friedman, J., Hastie, T., Tibshirani, R.: Additive logistic regression: a statistical view of boosting. Annals of Statistics 28(2), 337–407 (2000)
6. Friedman, S., Pasula, H., Fox, D.: Voronoi random fields: Extracting the topological structure of indoor environments via place labeling. In: Proceedings of the International Joint Conference on Artificial Intelligence, Hyderabad, India (2007)

7. Gonzalez, R.C., Wintz, P.A.: Digital Image Processing. Addison-Wesley Publishing Inc., Reading (1987)
8. Guyon, I., Elisseeff, A.: An introduction to variable and feature selection. Journal of Machine Learning Research 3, 1157–1182 (2003)
9. Haralick, R.M., Shapiro, L.G.: Computer and Robot Vision. Addison-Wesley Publishing Inc., Reading (1992)
10. Howard, A., Roy, N.: The robotics data set repository, Radish (2003), `http://radish.sourceforge.net/`
11. Kortenkamp, D., Weymouth, T.: Topological mapping for mobile robots using a combination of sonar and vision sensing. In: Proceedings of the National Conference on Artificial Intelligence, pp. 979–984 (1994)
12. Kuipers, B., Byun, Y.T.: A robot exploration and mapping strategy based on a semantic hierarchy of spatial representations. Journal of Robotics and Autonomous Systems 8, 47–63 (1991)
13. Loncaric, S.: A survey of shape analysis techniques. Pattern Recognition 31(8), 983–1001 (1998)
14. Montemerlo, M., Roy, N., Thrun, S.: Perspectives on standardization in mobile robot programming. In: Proceedings of the IEEE/RSJ International Conference on Intelligent Robots and Systems (2003)
15. O'Rourke, J.: Computational Geometry in C, 2nd edn. Cambridge University Press, Cambridge (1998)
16. Rosenfeld, A., Hummel, R.A., Zucker, S.W.: Scene labeling by relaxation operations. IEEE Transactions on Systems Man and Cybernetics 6(6), 420–433 (1976)
17. Rosenfeld, A., Pfaltz, J.L.: Sequential operations in digital picture processing. Journal of the Association for Computing Machinery 13(4), 471–494 (1966)
18. Russ, J.C.: The Image Processing Handbook. CRC Press, Boca Raton (1992)
19. Schapire, R.E., Singer, Y.: Improved boosting algorithms using confidence-rated predictions. Machine Learning 37(3), 297–336 (1999)
20. Shatkey, H., Kaelbling, L.P.: Learning topological maps with weak local odometric information. In: Proceedings of the International Conference on Artificial Intelligence (1997)
21. Tapus, A., Siegwart, R.: Incremental robot mapping with fingerprints of places. In: Proceedings of the IEEE/RSJ International Conference on Intelligent Robots and Systems, August 2005, pp. 2429–2434 (2005)
22. Taskar, B.: Learning associative markov networks. In: Proceedings of the International Conference on Machine Learning (2004)
23. Theodoridis, S., Koutroumbas, K.: Pattern Recognition, 3rd edn. Academic Press, London (2006)
24. Thrun, S.: Learning metric-topological maps for indoor mobile robot navigation. Artificial Intelligence 99(1), 21–71 (1998)
25. Triebel, R., Schmidt, R., Mozos, O.M., Burgard, W.: Instace-based AMN classification for improved object recognition in 2D and 3D laser range data. In: Proceedings of the International Joint Conference on Artificial Intelligence, Hyderabad, India, pp. 2225–2230 (2007)
26. Yamamoto, H.: A method of deriving compatibility coefficients for relaxation operators. Computer Graphics and Image Processing 10, 256–271 (1979)
27. Zivkovic, Z., Bakker, B., Kröse, B.: Hierarchical map building using visual landmarks and geometric constraints. In: Proceedings of the IEEE/RSJ International Conference on Intelligent Robots and Systems (2005)

Chapter 5
Probabilistic Semantic Classification of Trajectories*

5.1 Introduction

The approaches described in previous chapters are able to classify static observations using a mobile robot. However, mobile robots are dynamic agents that move along different trajectories. When operating in indoor environments, robots usually have a moderate velocity and a relatively continuous movement. That means, that observations obtained by a mobile robot at nearby poses are typically very similar. Furthermore, certain transitions between classes in a trajectory are rather unlikely. For example, if the classification of the current pose is *kitchen*, then it is rather unlikely that the classification of the next pose is *office* given the robot moved a short distance only. To get from the kitchen to the office, the robot first has to move through a doorway.

In this chapter, we present an approach that takes into account the dependencies between the classification of consecutive poses along a trajectory. In particular, we use a hidden Markov model (HMM) to filter the output of the current classification based on previous ones. In this way, we reduce the number of false classifications during the trajectory.

In addition, this chapter presents new places to be recognized in indoor environments. In particular, we want to recognize corridors, doorways, kitchens, seminar rooms, offices, and laboratories. For this purpose, we equip the robot with an additional camera and extract new features from images that permit us to extend the classification to the new places. The features extracted from the camera images are based on the recognition of objects. As an example, Fig. 5.1 shows an office environment together with some laser and vision data.

The complete approach first classifies the poses of the robot along a trajectory using a probabilistic decision list similar to the one introduced in Sect. 4.3. Then it applies a hidden Markov model to filter the current classification result based on previous ones. As a result, the mobile robot is able to classify the different places it traverses with high confidence.

* The work presented in this chapter originated from a collaboration with Axel Rottmann.

Ó.M. Mozos: Semantic Labeling of Places with Mobile Robots, STAR 61, pp. 57–69.
springerlink.com

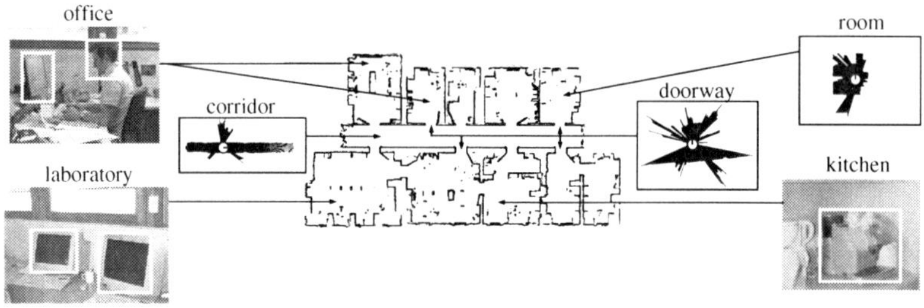

Fig. 5.1 An environment with offices, doorways, a corridor, a kitchen, and a laboratory. Additionally, the figure shows typical observations obtained by a mobile robot at different places.

The rest of the chapter is organized as follows. The next section introduces our modification of the AdaBoost algorithm to include the new weak classifiers for vision features. Section 5.3 describes the complete set of simple features extracted from laser and vision data. The models for the HMM are introduced in Sect. 5.4. In Sect. 5.5, experimental results obtained with this approach are presented. We discuss related work in Sect. 5.6. Finally, we conclude in Sect. 5.7.

5.2 Generalized AdaBoost

The generalized AdaBoost algorithm is a supervised learning algorithm designed to find a binary classifier that discriminates between positive and negative examples (cf. Sect. 2.2.2). AdaBoost boosts the classification performance of a simple learning algorithm by combining a collection of weak classifiers to a stronger one. The final strong classifier takes the form of a weighted combination of weak classifiers in which large weights are assigned to good classification functions whereas poor functions have small weights.

To classify the different places using laser and vision features, two types of weak classifiers are created. The first type is used for laser and vision features and has the form

$$h_j(x) = \begin{cases} +1 & \text{if } p_j f_j(x) < p_j \theta_j \\ -1 & \text{otherwise}, \end{cases} \tag{5.1}$$

where θ_j is a threshold, and p_j is either -1 or $+1$ and represents the direction of the inequality. Note that this form is the same as the one introduced in Sect. 4.2.

The second type of weak classifiers is designed for our set of vision features and has the form

$$h_j(x) = \begin{cases} p_j & \text{if } \theta_j^1 < f_j(x) < \theta_j^2 \\ -p_j & \text{otherwise}, \end{cases} \tag{5.2}$$

here θ_j^1 and θ_j^2 are thresholds delimiting an interval, and p_j is either $+1$ or -1 indicating whether the examples inside the interval are positive or negative. The form of this weak classifier is motivated by the fact that objects appear at different places in different numbers. For example, in an office room we expect more monitors than in the kitchen, but less than in the laboratory. Equation (5.2) is intended to encode this kind of information.

Finally, to learn the final strong classifier we use the algorithm introduced in Sect. 4.2. For the multiple class case, we use the same approach as in Sect. 4.3, and create a probabilistic decision list with binary classifiers.

5.3 Simple Features from Laser and Vision Data

In this chapter, we want to recognize an extended set of indoor places. This set is formed by corridors, doorways, kitchens, offices, laboratories, and seminar rooms. To recognize the additional places we equipped the robot with a laser scan and a camera. In the case of laser observations, the set of features are composed of the list presented in Sect. 3.4 and Sect. 4.4. These features are standard geometrical features used for shape recognition. Furthermore, they are rotational invariant to make the classification of a pose dependent only on the (x,y)-position of the robot and not on its orientation.

In the case of vision, the selection of the features is motivated by the fact that typical objects appear with different probabilities at different places. For example, the probability of detecting a computer monitor is larger in an office than in a kitchen. For each type of object, a vision feature is defined as a function that takes as argument a panoramic vision observation and returns the number of detected objects of this type in it. This number represents the single-valued feature f_j in the weak classifiers in (5.1) and (5.2).

In our case, we consider monitors, coffee machines, soap dispensers, office cupboards, frontal faces, face profiles, full human bodies, and upper human bodies. Example of such objects are shown in Fig. 5.1. The individual objects are detected using classifiers based on the set of Haar-like features proposed in [2].

5.4 Probabilistic Trajectory Classification

The approach described so far is able to classify single observations, but it does not take into account past classifications when determining the type of place at which the robot is currently located. However, whenever a mobile robot moves through an environment, the semantic labels of nearby places are typically identical. Furthermore, certain transitions between classes are unlikely. For example, if the robot is currently in a kitchen, then it is rather unlikely that the robot will end up in an office, given that it moved only a short distance. In many environments, to get from the kitchen to the office, the robot has to move through a doorway.

To incorporate such spatial dependencies between the individual classes, we apply a hidden Markov model and maintain a posterior belief $bel(y_t)$ about the type of place $y_t \in \mathrm{Y}$ the robot is currently at, where Y represents the set of possible semantic labels. The posterior is calculated as

$$bel(y_t) = \eta \, Pr(z_t \mid y_t) \sum_{y_{t-1}} Pr(y_t \mid y_{t-1}, u_{t-1}) bel(y_{t-1}). \tag{5.3}$$

In this equation, η is a normalizing constant ensuring that the left-hand side sums up to one over all classes y_t. To implement this HMM, three components need to be known. First, we need to specify the observation model $Pr(z_t \mid y_t)$, which is the likelihood of getting observation z_t given the actual class is y_t. In this case the observation z_t corresponds to the classification output of the probabilistic decision list. Second, we need to specify the transition model $Pr(y_t \mid y_{t-1}, u_{t-1})$, which defines the probability that the robot moves from class y_{t-1} to class y_t by executing action u_{t-1}. Finally, we need to specify how the belief $bel(y_0)$ is initialized.

In our current system, we choose a uniform distribution to initialize $bel(y_0)$. Furthermore, the classification output z_t is represented by a histogram P_t of probabilities over the set of all classes (cf. Sect. 4.3). In this histogram, the bin $p(k)$ stores the probability that the classified location belongs to the k-th class.

To determine $Pr(z_t \mid y_t)$, we use the KL-divergence between two histograms [1]. The first distribution is the current classification output $z_t = P_t$. The second one is learned from statistics: for each class y, we compute a histogram $\hat{z}_{1:m}(y)$ using m observations recorded within a place belonging to class y (here $m = 50$). This histogram $\hat{z}_{1:m}(y)$ is obtained by averaging out the individual histograms $\hat{z}_1, \ldots, \hat{z}_m$, which are computed according to (4.5). To determine $Pr(z_t \mid y_t)$, we use the KL-divergence $kld(\cdot \parallel \cdot)$ which provides a measure about the similarity of two distributions

$$Pr(z_t \mid y_t) = e^{-kld(z_t \parallel \hat{z}_{1:m}(y_t))} . \tag{5.4}$$

To illustrate the computation of the observation likelihood $Pr(z_t \mid y_t)$ consider Fig. 5.2. The first row depicts examples for the histograms $\hat{z}_{1:m}(y)$. The left image in the second row depicts the output z_t of the sequential classifier while the robot was in an office. As can be seen, also the classes doorway and seminar room have a probability significantly larger than zero. This output z_t and the histogram $\hat{z}_{1:m}(y_t)$ is then used to compute $Pr(z_t \mid y_t)$ according to (5.4). The result for all classes is depicted in the right image in the second row. In this image, each bin represents the likelihood $Pr(z_t \mid y_t)$ for the individual classes y_t. As can be seen, the observation likelihood given the robot is in a doorway is close to zero, whereas the likelihood given it is in an office is around 90%, which is actually the correct class.

To realize the transition model $Pr(y_t \mid y_{t-1}, u_{t-1})$, we only consider the two actions $u_{t-1} \in \{Move, Stay\}$. The transition probabilities were learned in a manually labeled environment by running 1000 simulation experiments. In each run, we started the robot at a randomly chosen point and orientation. We then executed a random movement so that the robot traveled between 20 cm and 50 cm forward. These values correspond to typical distances traveled by the robot between two consecutive updates of the HMM. The finally obtained transition probability matrix $Pr(y_t \mid y_{t-1}, u_{t-1})$ for the action *Move* is depicted in Fig. 5.3. As can be seen, the

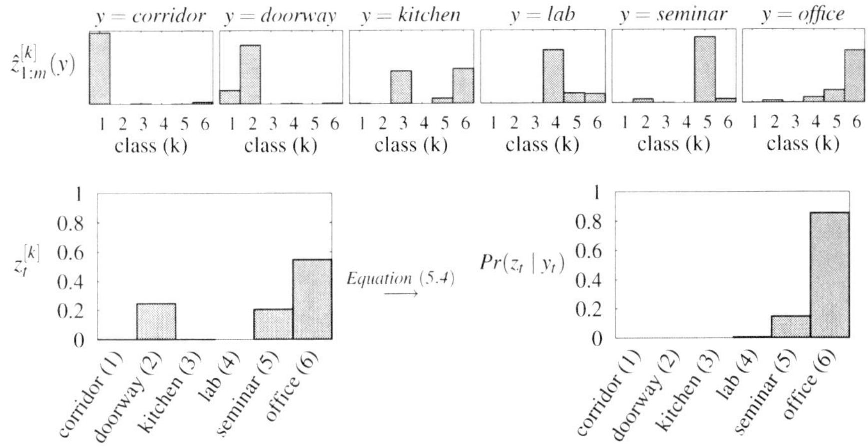

Fig. 5.2 The distributions depicted in the first row show the learned histograms $\hat{z}_{1:m}(y)$ for the individual classes: corridor (1), doorway (2), kitchen (3), lab (4), seminar room (5), and office (6). The left image in the second row depicts a possible classification output z_t. In the right image, each bar represents the corresponding likelihood $P(z_t \mid y_t)$ for the different estimates of y_t.

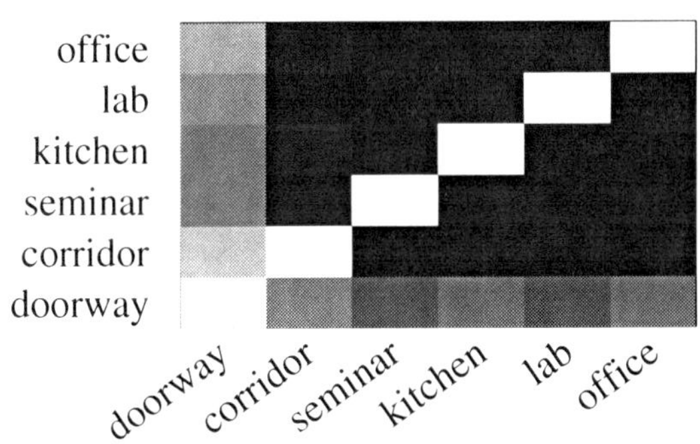

Fig. 5.3 The image depicts probabilities of possible transitions between places in the environment. To increase the visibility, we used a logarithmic scale. Dark values indicate low probability.

probability of staying in a place with the same classification is higher than the probability of changing the place. Moreover, the probability of moving from a room to a doorway is higher than the probability of moving from a room directly to a corridor. This indicates that the robot typically has to cross a doorway first in order to reach a different room. Furthermore, the matrix shows a lower probability of staying in a doorway than staying in the same type of room. This is due to the fact that a doorway is usually a small area in which the robot never rests for a longer period of time.

Table 5.1 Number T of weak classifiers and training error for the individual binary classifiers.

Binary Classifier	T	Training error [%]
lab	440	0.99
corridor	165	2.02
doorway	171	2.10
kitchen	68	2.46
seminar	334	2.58
office	288	7.31

5.5 Experimental Results

The approach described above has been implemented and tested using simulated and real robot data obtained in the office environment of building 79 at the University of Freiburg. The goal of the experiments is to demonstrate that the presented approach provides a robust classification of places in indoor environments into typical categories. We furthermore describe results indicating that the filtering of the classification output using an HMM significantly increases the performance of the overall approach. Additionally, we analyze the benefits of using vision features for the classification.

To train the classifier used throughout the experiments, we used 38,500 training examples. For each training example, we simulated the laser observations given an occupancy grid map of the environment. To generate the features extracted from vision data, we used 350 panoramic views recorded with our B21r robot, which is equipped with a SICK laser range finder and a camera system mounted on a pan/tilt unit as shown in Fig. 5.4. Each panoramic view consists of 8 images covering the 360° field of view around the robot. For each simulated laser scan, we then randomly drew a panoramic view from those corresponding to the type of the current place and used the vision features extracted from this view. As example, Fig. 5.5 shows two distributions over the number of coffee machines detected in the database images.

An important parameter of the AdaBoost algorithm is the number T of weak classifiers used to form the final strong binary classifier. For each strong binary classifier, we performed several experiments with up to 500 weak classifiers and analyzed the classification error. The number T of weak classifiers used to carry out the experiments has then been determined as the minimum in the error function. The resulting numbers T of weak classifiers used to form the strong binary classifiers and the classification errors of the finally obtained strong classifiers on the training data are given in Table 5.1.

In our current system, we determine the optimal sequence of strong binary classifiers by considering all possible sequences of strong binary classifiers. Although this approach is exponential in the number of classes, the actual number of permutations considered is limited in our domain due to the small number classes. In practice, we found out that the heuristic which sorts the classifiers in increasing order according to their training classification error also yields good results and at the same time can

Fig. 5.4 The image shows the robot Albert which was used for the experiments. Albert is a B21r robot equipped with a SICK laser range finder and a camera system mounted on a pan/tilt unit.

be computed efficiently. Compared to the optimal order, the classifier generated by this heuristic for an application with six different classes performed on average only 1.3% worse, as demonstrated in [6]. In several situations, the sequence generated by this heuristic turned out to be the optimal one.

5.5.1 Classifying Places along Trajectories

The first experiment is designed to demonstrate that the classifier learned from the training data in combination with the HMM can be used to robustly classify observation sequences acquired with a mobile robot in a real office environment. This environment contains six different types of places, namely offices, doorways, a laboratory, a kitchen, a seminar room, and a corridor. The ground truth for the different places in this environment is shown Fig. 5.6(a). We steered our robot through the environment and collected laser and image data along its trajectory. We then calculated the classification output without and with the HMM filtering and compared the results.

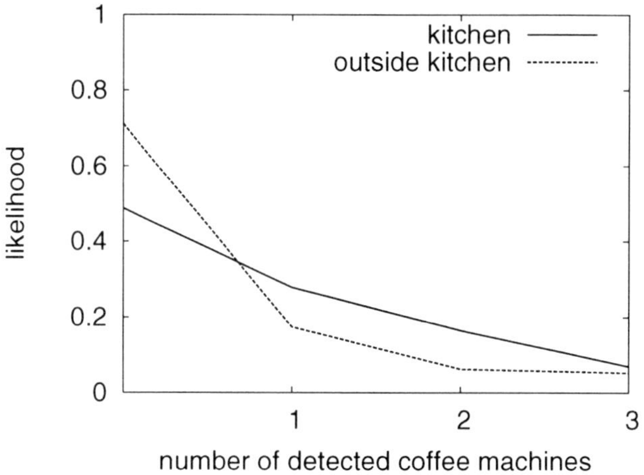

Fig. 5.5 Likelihood of detecting n coffee machines inside and outside a kitchen using Haar-like classifiers.

The classification rate of the sequential classifier without applying the HMM is 74.8%. The generated labels are shown in Fig. 5.6(b). If we additionally use the HMM to filter the output of the sequential classifier, the classification rate increases to 83.8%. The labels obtained after applying the HMM are shown in Fig. 5.6(c). As we can see in this example, the model for the HMM encodes the possible transitions and discards the ones with low probability. For example, the wrong office labels that appear in the kitchen (cf. Fig. 5.6(b)) are corrected after the application of the HMM (cf. Fig. 5.6(c)). The reason is that there is a very low probability of going directly from the kitchen to the office according to the model shown in Fig. 5.3. A two-sample t test revealed that the improvement of the resulting classification using an HMM is significant at the $\alpha = 0.01$ level. This illustrates that by using the HMM the overall classification rate can be improved seriously.

A second experiment was carried out using test data from a different part of the same building. We used the same sequential classifier as in the previous experiment. Whereas the sequential classifier yields a classification rate of 77.19%, the HMM generated the correct answer in 87.72% of all cases. This improvement is also significant at the $\alpha = 0.01$ level. Both results are shown in Fig. 5.7.

Finally, we studied how the HMM improves the final classification rate according to the output of AdaBoost. For this purpose, we analyzed the improvement of the HMM using different classification rates from AdaBoost. This is achieved by increasing the percentage of weak classifiers used in each binary classifier of the AdaBoost decision list. Here, 100% corresponds to the number of final weak classifiers used in the previous experiment (T in Table 5.1). For example, the classification rate decreases to 60% if only 5% of the final weak classifiers are used. The results are shown in Fig. 5.8. On average, the HMM improves the classification rate by 5.0%.

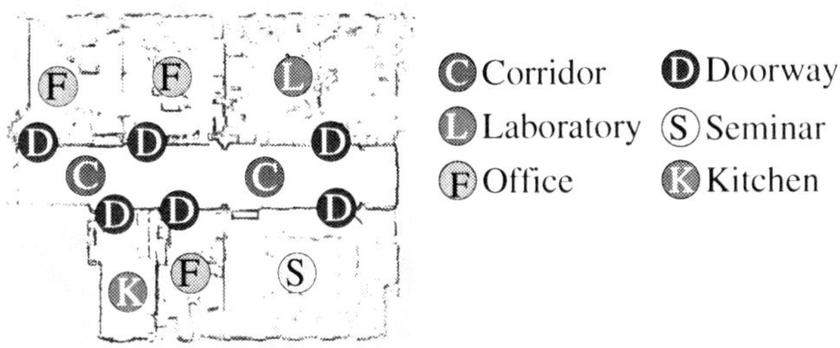

(a) Ground truth

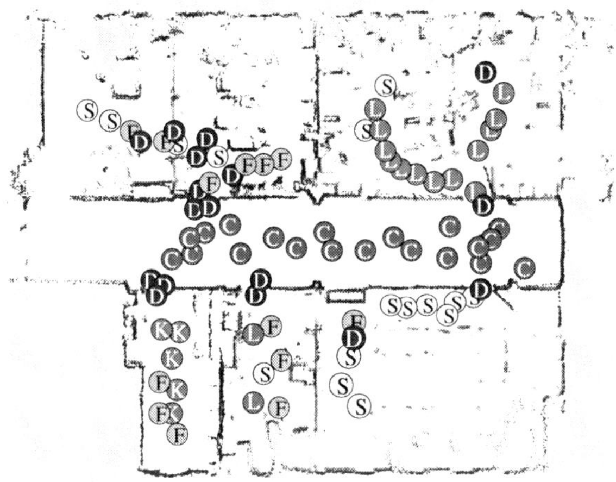

(b) AdaBoost classification

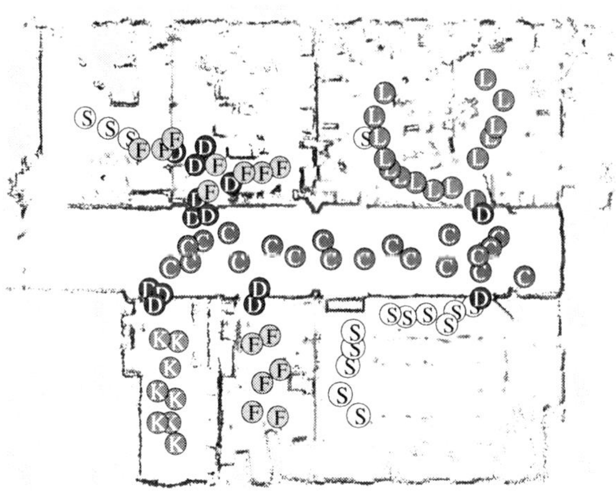

(c) HMM smoothing

Fig. 5.6 (**a**) Ground truth labeling of the individual areas in the environment. (**b**) Classification results obtained for a test set using only the output of the sequential classifier. (**c**) Smoothing of the classification applying the learned HMM.

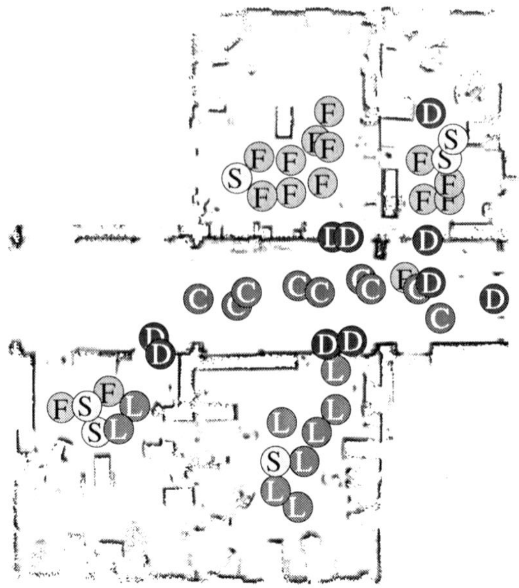

(a) AdaBoost classification

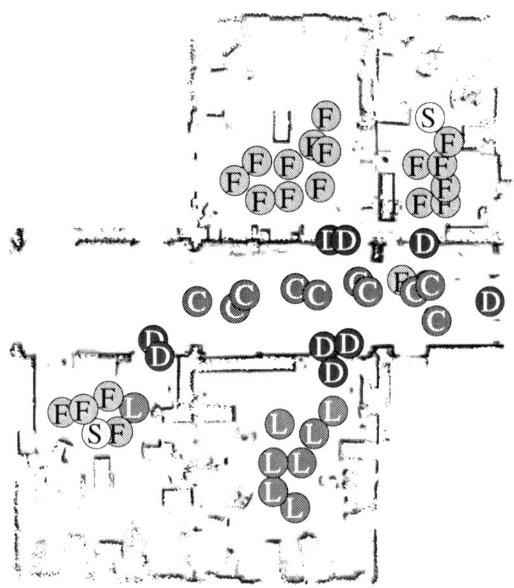

(b) HMM smoothing

Fig. 5.7 (**a**) Classification without filtering. (**b**) Classification using HMM smoothing.

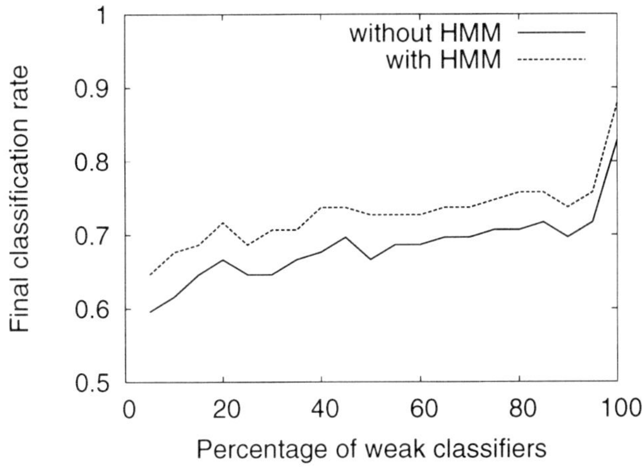

Fig. 5.8 Improvement of the HMM according to the percentage of weak classifiers used in each of the binary AdaBoost classifiers.

5.5.2 *Improvement Obtained by Combining Laser and Vision Data*

Additionally we analyzed whether the integration of vision and laser data yields any improvements over using only laser. To perform this experiment, we trained AdaBoost only with the three classes —office, corridor, and doorway—, because the other classes —kitchen, seminar room, and lab— can hardly be distinguished from offices using only laser observations. The classification errors obtained by integrating both modalities are summarized in Table 5.2. As can be seen, the combination of laser and vision data yields better results than the classifier relying on laser range data only.

Furthermore, some particular places can be hardly distinguished using only laser. A typical example can be found in the seminar and laboratory rooms. In our office environment, these places have a similar structure, and then similar observations

Table 5.2 Classification error obtained when using only laser data comparing to both laser and vision data.

Sequential Classifier	Error [%] laser	Error [%] laser & vision
corridor-doorway	3.21	1.87
doorway-room	3.74	2.67
doorway-corridor	3.21	2.14
room-corridor	1.60	1.34
corridor-room	1.60	1.34
room-doorway	1.60	1.60
average	2.50	1.83

are obtained using laser scans only. In these cases the addition of visual information improves the separability of the classes. For the seminar room, the error in the classification reduces from 46.9% to 6.3% when using additional vision data. When classifying the laboratory, the error decreases from 34.4% to 3.1%. This reduction in the classification indicates that some rooms are mainly distinguished by the objects found in them, like for example, monitors. These objects cannot be perceived with the laser sensor. A two-sample test indicates that the improvement is significant at the $\alpha = 0.01$ level.

5.6 Related Work

Classifying the places along a trajectory of a mobile robot is a relatively recent area of interest. One of the most known works is the one by Torralba et al. [8], which applies a hidden Markov model to distinguish between the places that a mobile robot traverses. Here, the information about the appearance of images is used to discriminate between different places. In contrast to this approach, the method presented in this chapter uses an additional laser range finder sensor. Moreover, we use the objects detected in the images instead of calculating visual features based on appearance. We classify the places based on their geometrical 2D structure and the objects found in them. In this way, we enable our robot to generalize better when classifying new environments.

Subsequent works analyze the capabilities for distinguishing places along a trajectory using camera images. Pronobis et al. [4] recognize the different places of an office environment using vision. Their approach is based on two kinds of features extracted from the images: interest points descriptors and appearance features. A similar approach is used by Luo et al. [3], but this time applying incremental learning. Also in the work by Spexard et al. [7], rooms are classified according to the appearance of images. In this case the goal of the robot is to recognize rooms previously seen. However, these approaches do not take into account past classifications when calculating the current semantic label.

In a recent study, Pronobis et al. [5] extend their previous work [4] with the additional use of a laser range finder and the set of geometrical features presented in this book. Results show that the laser features improve the generalization of the classifier. However, in [5] no HMM is used to smooth the classification.

For related work about semantic place classification of static poses we refer the reader to Sect. 3.7.

5.7 Conclusion

In this chapter, we have presented an approach to classifying the different poses along a trajectory into semantic classes. The technique uses a combination of simple geometric features extracted from laser range scans as well as features extracted from camera images. It further applies the AdaBoost algorithm to form a strong classifier. To distinguish between more than two classes, we use a sequence of binary classifiers arranged in a probabilistic decision list. To incorporate the spatial dependency between places, we apply a hidden Markov model that is updated upon sensory input and movements of the robot.

Our algorithm has been implemented and tested using a mobile robot equipped with a laser range finder and a camera system. Experiments carried out on a real robot as well as in simulation illustrate that our technique is well-suited to classifying trajectories in indoor environments. The experiments furthermore demonstrate that the hidden Markov model significantly improves the classification performance. Additional experiments revealed that the combination of vision and laser data increases the robustness and at the same time allows to distinguish between more classes compared to the approach in which only laser is used.

References

1. Cover, T.M., Thomas, J.A.: Elements of Information Theory. John Wiley & Sons, Chichester (1991)
2. Lienhart, R., Kuranov, A., Pisarevsky, V.: Empirical analysis of detection cascades of boosted classifiers for rapid object detection. In: Michaelis, B., Krell, G. (eds.) DAGM 2003. LNCS, vol. 2781, pp. 297–304. Springer, Heidelberg (2003)
3. Luo, J., Pronobis, A., Caputo, B., Jensfelt, P.: Incremental learning for place recognition in dynamic environments. In: Proceedings of the IEEE/RSJ International Conference on Intelligent Robots and Systems, San Diego, CA, USA (2007)
4. Pronobis, A., Caputo, B., Jensfelt, P., Christensen, H.I.: A discriminative approach to robust visual place recognition. In: Proceedings of the IEEE/RSJ International Conference on Intelligent Robots and Systems, Beijing, China (2006)
5. Pronobis, A., Mozos, O.M., Caputo, B.: SVM-based discriminative accumulation scheme for place recognition. In: Proceedings of the IEEE International Conference on Robotics and Automation, Pasadena, California, USA (2008)
6. Rottmann, A.: Bild- und laserbasierte klassifikation von umgebungen mit mobilen robotern. Master's thesis, University of Freiburg, Department of Computer Science (2005) (in German)
7. Spexard, T., Li, S., Wrede, B., Fritsch, J., Sagerer, G., Booij, O., Zivkovic, Z., Terwijn, B., Kröse, B.: BIRON, where are you? - enabling a robot to learn new places in a real home environment by integrating spoken dialog and visual localization. In: Proceedings of the IEEE/RSJ International Conference on Intelligent Robots and Systems (2006)
8. Torralba, A., Murphy, K.P., Freeman, W.T., Rubin, M.A.: Context-based vision system for place and object recognition. In: Proceedings of the International Conference on Computer Vision (2003)

Chapter 6
Semantic Information in Exploration and Localization*

6.1 Introduction

The work presented in the previous chapters showed how to augment the representation of indoor environments using semantic information about places. In this chapter we describe how robots can use the intrinsic information of human-made environments to improve their actions. In particular, we apply the semantic labeling of places to two robotic tasks: multi-robot exploration, and localization. In both cases the performance of the robot increases when it takes into account the classification of its location.

Exploration and localization belong to the fundamental problems in mobile robotics [22]. In the exploration task a mobile robot is controlled in a way that maximizes the information about its environment. A typical goal in exploration consists of creating a map of a previously unseen environment. Moreover, the use of multiple robots is often suggested to have advantages over a single robot during exploration, since cooperating robots have the potential to accomplish a task faster than a single robot [7]. In the localization task, a mobile robot has to determine its pose relative to a map of a given environment.

In this chapter, we first present an approach to include semantic information about places to better distribute the robots in human-made environments during an exploration task. As we have seen in previous chapters, indoor environments constructed by humans contain structures like corridors, rooms or offices. Moreover, corridors are connected to several rooms and provide more branchings to new unexplored areas. The key idea is then to assign higher rewards to robots that first explore corridors. As a result, the overall completion time of an exploration task can be significantly reduced.

In a second approach, we use the semantic labeling to localize a robot in an indoor environment using the Monte Carlo localization approach [3]. The main idea here is to take as observation model the semantic classification of the current pose of the mobile robot.

---

* The work presented in this chapter originated from a collaboration with Cyrill Stachniss.

Ó.M. Mozos: Semantic Labeling of Places with Mobile Robots, STAR 61, pp. 71–81.
springerlink.com

The rest of the chapter is organized as follows. The next section presents the approach for multi-robot exploration using semantic information. In Sect. 6.3, we introduce the Monte Carlo approach for localization using semantic labels. Section 6.4 presents experimental results. We discuss related work in Sect. 6.5. Finally, conclusions are presented in Sect. 6.6.

6.2 Multi-robot Exploration Using Semantic Information

In multi-robot exploration, a team of robots is distributed in a new environment with the objective of accumulating information to create a map. In this task, we are interested in finding good assignments of goal positions for the robots in the team. In our case, we assign target locations with the aim of minimizing the time needed to complete the exploration.

6.2.1 Classifying Target Locations

We assume that the knowledge about the environment is represented by an occupancy grid map. In this representation, target locations are found at the frontier between known and unknown areas [24]. As an example, Fig. 6.1(a) shows a map together with the frontiers detected there (dashed lines). For each of the frontiers, a target location is generated.

The goal now is to classify each potential target location into a semantic class. One possible solution to classify a target location is to simulate an observation at its position, and then classify this observation using the approach presented in Chap. 3. However, the target position is located at a frontier, which means that part of the neighboring areas are not known. Therefore, the laser observations simulated at frontier cells contain a significant number of maximum-range readings, which can lead to high missclassification rates.

To increase the classification rate in these cases, a short virtual trajectory to the desired goal location is generated. We then simulate laser range observations at different poses along this virtual trajectory using the partially known map. These poses

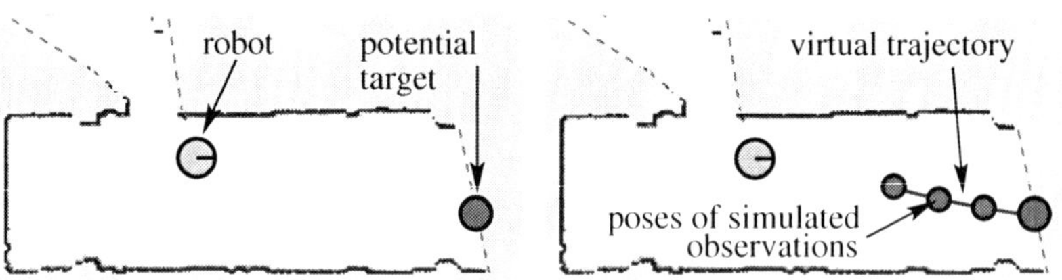

(a) Frontier extraction (b) Virtual trajectory

Fig. 6.1 (**a**) The image shows a situation in which a robot has extracted the frontiers of the occupancy grid map (*dashed lines*). Additionally, a target location is shown for one of the frontiers. (**b**) A virtual trajectory to the target is generated by the robot.

are generated selecting cells in the occupancy grid which are as far away as possible from the unknown locations in the current map using the euclidean distance transformation [14]. The reason for this selection is that cells having more information about its surroundings will have a lower error in their semantic classification, since their simulated range scans will contain fewer maximum-readings. Then an A* planner is used to generate the virtual trajectory to the target location. An example illustrating this process is shown in Fig. 6.1(b).

Once the virtual trajectory is created, we follow the approach presented in Chap. 5 to classify trajectories. This method applies a hidden Markov model (HMM) to maintain a posterior $bel(t_t)$ about the type y_t of the place the virtual sensor is located at as

$$bel(y_t) = \eta\, Pr(z_t \mid y_t) \sum_{y_{t-1}} Pr(y_t \mid y_{t-1}, u_{t-1}) bel(y_{t-1}). \tag{6.1}$$

The different components of this model are calculated in the same way as explained in Chap. 5. Using (6.1), we classify the target location $bel(y_{\text{target}})$ using the classification of the positions leading to it.

6.2.2 Target Assignment Using Semantic Place Labeling

As indicated above, the main idea is to give priority to target locations that are located in corridors, because they lead to a higher number of unknown areas. This is achieved by using the algorithm used for target assignment shown in Fig. 6.2.

- Determine the set of frontier cells.
- Compute for each robot i the cost V_t^i for reaching each frontier cell t.
- Assign to each frontier cell t a semantic labeling L_t.
- Set the utility U_t of all frontier cells t to $U_{init}(L_t, n)$ according to their semantic labeling L_t and the size n of the team.
- While there is one robot left without a target point

  1. Determine a robot i and a frontier cell t which satisfy

  $$(i,t) = \underset{(i',t')}{\operatorname{argmax}} \left(U_{t'} - V_{t'}^{i'} \right).$$

  2. Reduce the utility of each target point t' in the visibility area according to

  $$U_{t'} \leftarrow U_{t'} - Pr_{vis}(t,t').$$

Fig. 6.2 The algorithm for the assignation of target locations to the different robots in a team.

The algorithm first calculates the set of frontier cells in the current submap. For each robot i in the team, the algorithm then calculates the cost V_t^i of reaching each frontier cell t. This cost is based on the distance the robot has to travel to reach the cell. Additionally, the algorithm estimates the semantic label y_t of the target location t using the HMM-based approach presented in Sect. 6.2.1.

Using the label y_t together with the number n of robots in a team, an initial utility function $U_{init}(y_t, n)$ is assigned to each target location t. In this function, the target locations classified as corridors get an initial utility U_{init} which is β times higher than other target locations. In our implementation, we selected a value of $\beta = 5$, since after several experiments we found that this value led to the best results in different runs of the algorithm.

The algorithm continues with an iterative process which combines two steps. First, the best combination of robot i and target t is selected. This selection is done maximizing the utility function U_t. A frontier cell is discounted as soon as it is selected. In this way, we avoid the situation in which several robots received the same frontier cell. We additionally discount target locations which can potentially be observed by other robots that already have a target assigned. This is done by using the utility function

$$U(t_n \mid t_1, \ldots, t_{n-1}) = U_{t_n} - \sum_{i=1}^{n-1} Pr_{vis}(t_n, t_i) \,, \tag{6.2}$$

with $Pr_{vis}(t_n, t_i)$ being the probability that the frontier t_n can be observed by a robot moving to frontier cell t_i. We approximate this probability density using a linear function.

The algorithm of Fig. 6.2 reduces the interference of robots during the exploration taking into account the visibility constraints, which are included in the utility function. Moreover, the inclusion of semantic information about the target locations improves the distribution of robots, giving preference to corridor places when selecting goal position for exploring unknown areas. As a result, the robots are better distributed and the time needed to explore the environment is significantly reduced.

The reduction of time during the exploration is not significant when the number of members in the team is small. This fact can be explained by considering the single-robot exploration scenario. To create a map of the environment, a single robot has to explore the whole environment, and it makes no sense to focus first on a particular place like a corridor. In our experiments, the exploration time does not decrease if the team has less than five robots.

6.3 Localization Using Place Recognition

A second problem which benefits from the semantic classification of places is the localization of a robot in an indoor environment. In this task the robot is given a representation of an indoor environment in the form of an occupancy grid map [4, 15]. Each cell in this grid additionally contains a semantic label about its place.

In our case the set of possible places to be recognized is composed of a corridor, doorways, offices, a kitchen, a seminar room, and a laboratory. This set of places is similar to the one used in Chap. 5.

The method applied to localize the robot in the environment is the popular Monte Carlo Localization (MCL) approach [3]. This localization method applies a recursive Bayesian scheme to maintain a posterior about the location of the robot x_t given the map m of the environment, the odometry information $u_{0:t-1}$, and the observations $z_{1:t}$ as

$$bel(x_t \mid m, z_{1:t}, u_{0:t-1}) = \eta \cdot Pr(z_t \mid m, x_t) \cdot Pr(x_t \mid m) \cdot \qquad (6.3)$$
$$\cdot \int_{x'} Pr(x_t \mid x', u_{t-1}) \cdot bel(x' \mid m, z_{1:t-1}, u_{0:t-2})\, dx'.$$

In MCL, the posterior about the robot positions is estimated using particle filters. The belief $bel(x_t)$ is represented by a set of random samples or particles in a non parametric form. This representation can approximate a broad set of distributions [22].

Our implementation of MCL is characterized by the observation model which uses place semantic information. As observations $z_{1:t}$, we use the output of the classifier the robot uses for place labeling. This classifier is the same as the one introduced in Sect. 5.2, and applies a probabilistic decision list in which each element contains a binary classifier. The quantity $Pr(z_t \mid m, x_t)$ is then determined as $Pr(z_t \mid y_t)$, where y_t is the class assigned to x_t in m. To estimate $Pr(z_t \mid y_t)$, we generated statistics about the output of the sequential multi-class classifier given the robot was at a place corresponding to y_t. Additionally, we weight the particles inversely proportional to the occupancy probability at x_t in m.

6.4 Experimental Results

To show the improvements obtained in the exploration and localization tasks when using semantic information about places we carried out several experiments using simulation and real robots. The approach for multi-robot exploration was implemented using teams with different number of robots. Due to the high numbers of robots used, we evaluated our technique only in simulation experiments. For the robot localization we used an ActivMedia Pioneer II robot with two lasers.

6.4.1 Multi-robot Exploration

The following experiments were designed to show the improvements of the multi-robot exploration technique making use of semantic place information. We evaluate the approach on simulation due to the large number of robots that form the teams. Further evaluations can be found in [21].

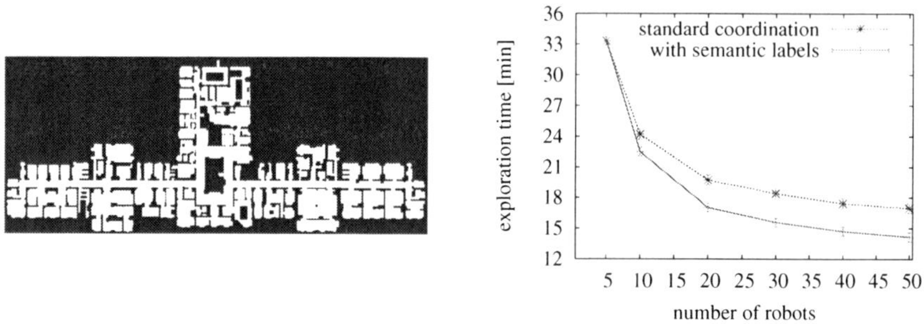

(a) Fort Sam Huston hospital

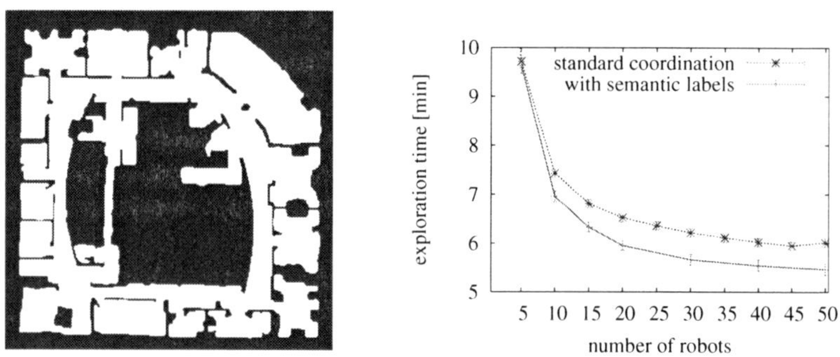

(b) Intel Research Lab

Fig. 6.3 (**a**) Map of the Fort Sam Huston hospital, and performance when the semantic information is taken into account in comparison to the case where no label information is used. (**b**) Map of the Intel research Lab with its corresponding performance plot.

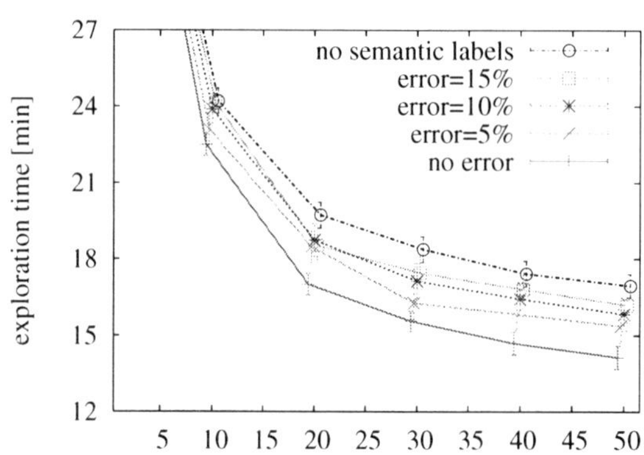

Fig. 6.4 Exploration performance with different classification errors.

In a first experiment we used the map of the Fort Sam Huston hospital (cf. Fig. 6.3(a)), which contains several corridors together with rooms adjacent to them. In the experiment we apply our method for coordinating several robots using

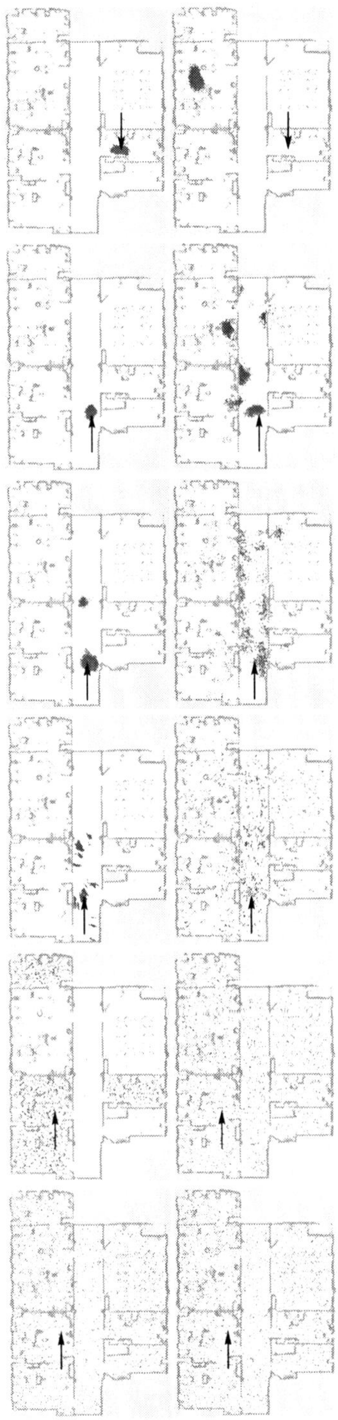

Fig. 6.5 Global localization using semantic information and odometry information (first row) compared to an approach using only the odometry information (second row). The images in the same column represent the corresponding filter at the same time. The arrow indicates the ground truth position. As the results indicate, semantic information can be used to speed up global localization.

semantic information about places, and compared it to the case in which no place information is used. In addition, Fig. 6.3(a) shows the performance when the number of robots in the team varies from 5 to 50 members. For each team size, we repeated the experiments 50 times. In all the experiments the robots started from the same initial position. As the plot shows, the time needed to explore the complete environment is significantly reduced at the confidence level of 0.05 when using semantic place information. A similar experiment was carried out using the map of the Intel Research Lab. Again we observed a significant reduction in the exploration time as shown in Fig. 6.3(b).

In the previous experiments we assumed that the semantic classification of the target locations had no errors. In real situations, however, errors usually appear during the labeling process (see experimental results in Chaps. 3 and 5). To study the performance of the method according to the classification errors, we carried out an experiment in which we randomly misclassified different percentages of target locations and measure the exploration time according to them. Figure 6.4 shows the resulting performance using different team sizes. When the error in the classification exceeds 15% the improvement using semantic information is not significant anymore.

6.4.2 *Localization*

The last experiment is designed to illustrate how semantic information about places can be used to improve the localization of a mobile robot in its environment. In this experiment, we used an ActivMedia Pioneer II robot equipped with two laser range finders covering 360° around the robot (cf. Fig 3.8(a)). Note that the laser data is only fed into the semantic classifier and not used for metric localization.

We apply a Monte Carlo localization approach following the model in (6.3). Figure 6.5 illustrates the evolution of two particle sets over time. In the first row, the semantic information was available whereas in the second row only the odometry information was used. Both filters were initialized with a uniform distribution with $10,000$ particles. The robot initially was located in the second left office, north of the corridor. Therefore, particles located in offices received higher importance weights compared to the other samples. Whereas the approach utilizing semantic information converges quickly to the correct solution, the particle filter that relies only on the odometry information $Pr(x_t \mid m)$ finally diverges.

6.5 Related Work

Different aspects of multi-robot exploration have been studied in the past. For example, Yamauchi [23] presented a technique to learn maps with a team of mobile robots. In this approach, the robots exchange information about the map that is continuously updated whenever new sensor input arrives. Koenig et al. [12] analyze different terrain coverage methods for simple robots with limited sensing and computational capabilities.

In the work by Zlot et al. [25], a technique is presented in which the exploration is guided by a market economy. It considers sequences of potential target locations for each robot and tasks are traded between the robots using single-item first-price sealed-bid auctions. Singh and Fujimura [20] present a method for heterogeneous robot teams. In this approach, if a robot is too big to pass through a narrow passage, it informs other robots to do this task. Howard et al. [9] introduce an incremental deployment approach that explicitly deals with situations in which the path of one robot is blocked by another. Also Matarić and Sukhatme [13] introduce different strategies for allocating tasks in robot teams, and analyze their performance in different experiments. Finally, the Hungarian method to compute the assignments of frontier cells to robots is introduced by Ko et al. [11].

The coordination technique presented is this chapter is an extension of the work by Burgard et al. [2]. We also discount the utility of target locations if they are visible from a goal location already assigned to a robot. In contrast to [2], our approach estimates and incorporates background knowledge about environmental structure into the goal point assignment procedure.

The semantic labels used to improve multi-robot coordination can be seen as background knowledge about spacial structures. Fox et al. [6] presented a technique which learns background knowledge in typical indoor environments and later on use that knowledge for map building. They apply their approach to decide whether the robot is seeing a previously built portion of a map, or is exploring new terrain.

Localization is a typical problem in mobile robotics, and different approaches have been applied to solve it. The grid-based Monte Carlo localization was introduced by Simmons and Koenig [18]. This approach approximates the posterior of the robot pose using a histogram over the possible discrete poses. Several authors have successfully applied grid-based Monte Carlo localization in their work, as for example Burgard et al. [1], Hertzberg and Kirchner [8], and Simmons et al. [19]. Multi-hypothesis extended Kalman filters is another approach for localization used by different authors, as for instance Jensfelt and Kristensen [10], Roumeliotis and Bekey [17], and Reuter [16]. Finally, particle filter approaches were introduced by Dellaert et al. [3] and Fox et al. [5]. In this chapter we use this approach but additionally include the semantic classification of places.

For related work about the semantic labeling of places we refer the reader to Sects. 3.7 and 5.6.

6.6 Conclusion

In this chapter, we have shown how the semantic information helps to improve several robotic tasks. In particular, we proposed a technique that takes into account semantic information about places in the context of coordinated multi-robot exploration. The main idea is that mobile robots can use the intrinsic information of human-made environments to improve their actions. This improvement is obtained by selecting the best target locations according to their semantic classification. The semantic labeling of the target locations is done using an AdaBoost-based classifier.

Additionally, a hidden Markov model is applied to improve the classification in a virtual trajectory to the target position.

Alternatively we have seen how the semantic information about places can be used to localize the robot in an indoor environment using the Monte Carlo localization approach. In this case, the observation model of the robot corresponds to the semantic classification of its position. Experimental results indicate that this approach can be used to speed up global localization.

Both methods demonstrated that the semantic information can improve the performance in different tasks using autonomous mobile robots.

References

1. Burgard, W., Cremers, A.B., Fox, D., Hähnel, D., Lakemeyer, G., Schulz, D., Steiner, W., Thrun, S.: Experiences with an interactive museum tour-guide robot. Artificial Intelligence 114(1-2) (2000)
2. Burgard, W., Moors, M., Stacniss, C., Schneider, F.: Coordinated multi-robot exploration. IEEE Transactions on Robotics 21(3), 376–378 (2005)
3. Dellaert, F., Fox, D., Burgard, W., Thrun, S.: Monte carlo localization for mobile robots. In: Proceedings of the IEEE International Conference on Robotics and Automation (1999)
4. Elfes, A.: Using occupancy grids for mobile robot perception and navigation. Computer 22(6), 46–57 (1989)
5. Fox, D., Burgard, W., Dellaert, F., Thrun, S.: Monte carlo localization: Efficient position estimation for mobile robots. In: Proc. of the National Conference on Artificial Intelligence (1999)
6. Fox, D., Ko, J., Konolige, K., Stewart, B.: A hierarchical bayesian approach to the revisiting problem in mobile robot map building. In: Proceedings of the International Symposium of Robotics Research, Siena, Italy (2003)
7. Guzzoni, D., Cheyer, A., Julia, L., Konolige, K.: Many robots make short work. AI Magazine 18(1), 55–64 (1997)
8. Hertzberg, J., Kirchner, F.: Landmark-based autonomous navigation in sewerage pipes. In: Proceedings of the First Euromicro Workshop on Advanced Mobile Robots (1996)
9. Howard, A., Mataric, M.J., Sukhatme, G.S.: An incremental deployment algorithm for mobile robot teams. In: Proceedings of the IEEE/RSJ International Conference on Intelligent Robots and Systems, Lausanne, Switzerland, pp. 2849–2854 (2002)
10. Jensfelt, P., Kristensen, S.: Active global localisation for a mobile robot using multiple hypothesis tracking. IEEE Transactions on Robotics and Automation 17(5), 748–760 (2001)
11. Ko, J., Stewart, B., Fox, D., Konolige, K., Limketkai, B.: A practical, decision-theoretic approach to multi-robot mapping and exploration. In: Proceedings of the IEEE/RSJ International Conference on Intelligent Robots and Systems, Las Vegas, NV, USA, pp. 3232–3238 (2003)
12. Koenig, S., Szymanski, B., Liu, Y.: Efficient and inefficient ant coverage methods. Annals of Mathematics and Artificial Intelligence 31, 41–76 (2001)
13. Matarić, M.J., Sukhatme, G.S.: Task-allocation and coordination of multiple robots for planetary exploration. In: Proceedings of the International Conference on Advanced Robotics, Budapest, Hungary, pp. 61–70 (2001)

14. Meijster, A., Roerdink, J.B.T.M., Hesselink, W.H.: A general algorithm for computing distance transforms in linear time. In: Mathematical Morphology and its Applications to Image and Signal Processing, pp. 331–340. Kluwer Academic Publishers, Dordrecht (2000)
15. Moravec, H.P.: Sensor fusion in certainty grids for mobile robots. AI Magazine 9, 61–74 (1988)
16. Reuter, J.: Mobile robot self-localization using PDAB. In: Proceedings of the IEEE International Conference on Robotics and Automation, San Francisco, CA, USA, pp. 3512–3518 (2000)
17. Roumeliotis, S.I., Bekey, G.A.: Bayesian estimation and kalman filtering: a unified framework for mobile robot localization. In: Proceedings of the IEEE International Conference on Robotics and Automation, San Francisco, CA, USA, pp. 2985–2992 (2000)
18. Simmons, R., Koenig, S.: Probabilistic robot navigation in partially observable environments. In: Proceedings of the International Joint Conference on Artificial Intelligence, pp. 1080–1087 (1995)
19. Simmons, R., Lopez Fernandez, J., Goodwin, R., Koenig, S., O'Sullivan, J.: Lessons learned from Xavier. Robotics and Automation Magazine, 733–739 (2000)
20. Singh, K., Fujimura, K.: Map making by cooperating mobile robots. In: Proceedings of the IEEE International Conference on Robotics and Automation, Atlanta, GA, USA, pp. 254–259 (1993)
21. Stachniss, C., Mozos, O.M., Burgard, W.: Speeding-up multi-robot exploration by considering semantic place information. In: Proceedings of the IEEE International Conference on Robotics and Automation (ICRA), Orlando, FL, USA, pp. 1692–1697 (2006)
22. Thrun, S., Burgard, W., Fox, D.: Probabilistic Robotics. MIT Press, Cambridge (2005)
23. Yamauchi, B.: Frontier-based exploration using multiple robots. In: Proceedings of the International Conference on Autonomous Agents, Minneapolis, MN, USA, pp. 47–53 (1998)
24. Yamauchi, B., Schultz, A., Adams, W.: Integrating exploration and localization for mobile robots. Adaptive Behavior 7(2), 217–229 (1999)
25. Zlot, R., Stentz, A., Dias, M.B., Thayer, S.: Multi-robot exploration controlled by a market economy. In: Proceedings of the IEEE International Conference on Robotics and Automation, Washington, DC, USA (2002)

Chapter 7
Conceptual Spatial Representation of Indoor Environments*

7.1 Introduction

In the last years, there has been an increasing interest in service robots, such as domestic or elderly care robots, whose purpose is to assist people in human-like environments. These service robots have to interact with people having little or no formal training in robotics. In such situations, the communication and interaction between robots and humans become key issues for these systems.

One of the most intuitive and powerful ways for humans to communicate is spoken language. It is therefore interesting to design robots that are able to speak with people and understand their words and expressions. For this, the robot needs to perceive the world in a similar way humans do. However, when comparing the way robots typically perceive and represent the world with the findings from cognitive psychology on how humans do it, it is evident that there is a large discrepancy. Bridging the gap between human and robot spatial representations is thus of great importance.

In this chapter we give an overview of an integrated approach for creating conceptual representations of human-made environments using a mobile robot. In this representation, the concepts refer to spatial and functional properties of typical places found in indoor environments. The complete model is composed of layers containing maps at different levels of abstraction. The lower layers contain a metric map, a navigation map and a topological map, each of which plays a role in navigation and self-localization of the robot. On the topmost level of abstraction, the conceptual map provides a richer semantic view of the spatial organization.

The complete multi-layered representation is created using a combination of user-driven map acquisition process together with autonomous exploration and discovery of the environment. This process is actively supported by a linguistic framework, which allows the user to communicate with the robot using natural language only. The complete system in shown in Fig.7.1.

The rest of the chapter is organized as follows. In the next section, we present the multi-layered conceptual spatial representation. In Sect. 7.3, we describe

---

* This chapter originated from a joint work with Hendrik Zender and Patric Jensfelt.

Ó.M. Mozos: Semantic Labeling of Places with Mobile Robots, STAR 61, pp. 83–97.
springerlink.com

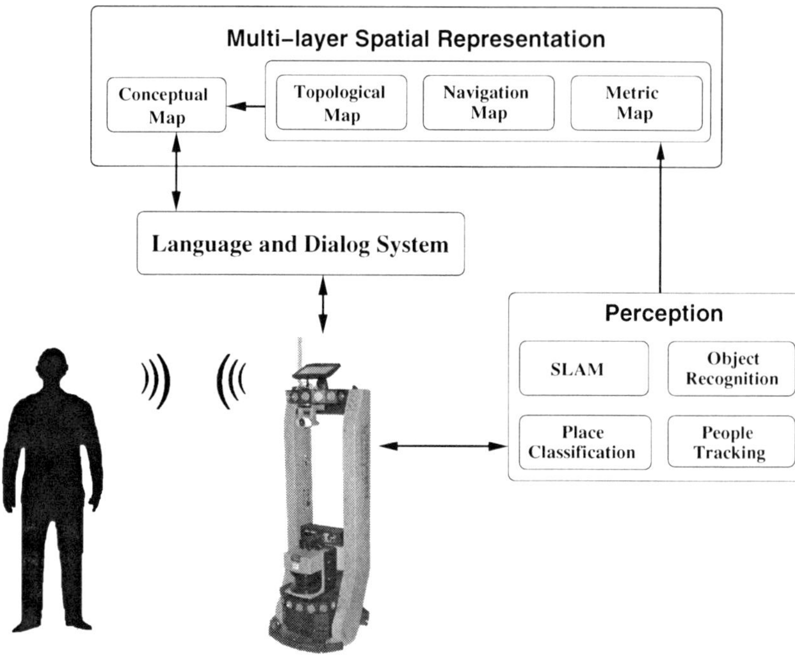

Fig. 7.1 The integrated system used for conceptual spatial representations of indoor environments. The arrows indicate the direction in which the information flows along the different subsystems. The communication between the user and the robot is done using natural language only.

implementation details of the complete system. Section 7.4 presents a demo which shows the capabilities of the service robot. We discuss related work in Sect. 7.5. Finally, we conclude in Sect. 7.6.

7.2 Multi-layered Conceptual Mapping

The goal of the multi-layered conceptual mapping is to generate spatial representations that enable a mobile robot to create models of human-made environments in a similar way as people do. The concepts represented in this model correspond to spatial and functional properties of typical indoor environments.

Our representation is based on findings in cognitive psychology [15] that assume that topological areas are the basic spatial units suitable for situated interaction between humans and robots. In addition, people usually refer to places according to the functions ascribed to them.

Taking these ideas into account, the final representation model is divided into layers, each representing a different level of abstraction. At the lowest level the system uses a laser scanner and the odometry of the mobile robot to create a metric map of the environment. On top of the metric representation, a navigation map is constructed. The navigation map is used by the mobile robot to travel along routes. In an upper layer we find the topological map, which represents the different areas found in the environment. This layer uses the detected doorways as criteria for splitting areas in the environment. On top of these layers the conceptual map contains the spatial information and the knowledge about the objects found in the environment, as well as the relations between them. This layer gathers information coming from lower maps together with information from different modalities such as proximity information, vision or dialogue, to allow symbolic reasoning and situated dialogue. Fig. 7.2 depicts the four layers of the conceptual spatial representation.

7.2.1 Metric Map

The first layer of the model contains a metric representation of the environment in an absolute frame of reference (cf. Fig. 7.2, bottom). The geometric primitives of this metric map consist of lines extracted from laser range scans. Such lines typically correspond to walls and other flat structures in the environment. The complete metric map is created by a mobile robot using simultaneous localization and mapping (SLAM) techniques. In particular, we apply the framework introduced in [5], which uses general representations for features that address symmetries and constraints in the feature coordinates to be added to the map with partial initialization. The number of dimensions for a feature can grow with time as more information is acquired. The basis for integrating the feature observations is the extended Kalman filter (EKF). An example metric map created using this method is shown in Fig. 7.3.

7.2.2 Navigation Map

The second layer contains the navigation map, which is represented by a graph (cf. Fig. 7.2). This representation is based on the notion of a roadmap of virtual free-space markers [13, 16]. In this approach, navigation nodes are inserted in the map as the robot moves through the environment. A new node is added whenever the robot has traveled a certain distance from the closest existing one. The graph serves for planning and autonomous navigation in the known part of the environment.

The navigation nodes are classified into *doorway* and *place* nodes. Doorway nodes indicate the transition between different places and represent possible doors. They are detected and added whenever the robot passes through a narrow opening. Later, the status (open/closed) of a known door can be monitored using the laser scanner. Additionally, doorway nodes are assigned information about the door opening such as its width and orientation.

Place nodes are in turn classified into two classes, namely *corridor* and *room*. The classification of place nodes is done following the approach introduced in Chap. 3. Laser observations are constantly obtained by the robot and classified into one of the two previous classes. When the robot is located at a position in the map corresponding to a place node, it assigns this node the label resulting from the classification of the current observation. To increase the robustness of this method the robot constantly

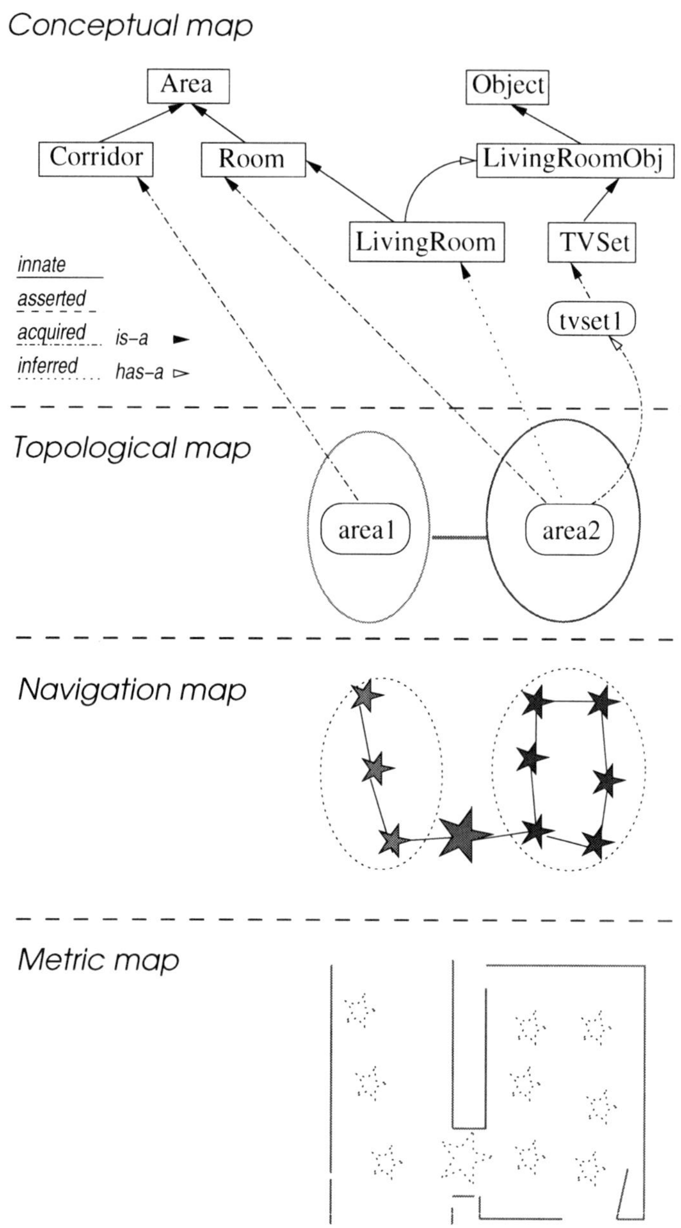

Fig. 7.2 Layered spatial representation for an indoor environment.

stores the last N classifications. Whenever a place node is reached, a majority vote is obtained with these last N labels and a final classification is assigned to the node.

The place classification of Chap. 3 is based on a 360° field of view. However the robot used in our system is equipped only with one laser covering a restricted 180°

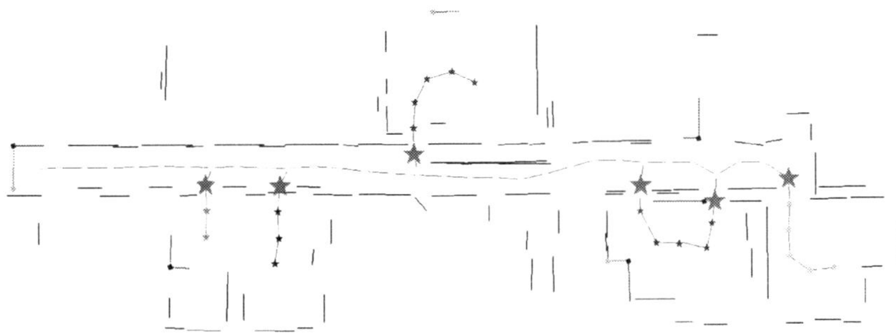

Fig. 7.3 Example of the two first layers in the spatial representation. The metric map is composed of lines. The navigation map is represented by stars. Different areas are represented by different colors/gray levels. Big stars indicate the doorways separating the places.

field of view at the front of the robot (see the robot in Fig 7.1). To solve this problem we follow the approach described in Sect. 3.5 and maintain a local map around the robot which permits us to simulate the rest of the beams covering the rear part of the robot. The classification of places and doorways forms the basis of the conceptual system.

7.2.3 Topological Map

The topological map divides the set of nodes in the navigation graph into different areas. An area consists of a set of interconnected nodes with the same place classification. The nodes are partitioned on the basis of the detection of doorways. This process is shown in Fig. 7.2.

In the topological layer, the exact shape and boundaries of an area are irrelevant. This approach complies with previous studies [8, 15], which state that humans segment space into regions that correspond to more or less clearly defined spatial areas.

Note that this method for topological map extraction is an alternative to the one presented in Chap. 4, where we applied an offline approach using simulated range data for the classification of the free poses in the map. Moreover, that method needed the complete map before extracting the topology. In contrast, the approach of this chapter is based mainly on the detection of doorways as the boundaries between different regions. Then the nodes in the different regions are labeled according to their semantic classification. This procedure is more appropriate for an online creation of the topological map during a guided tour.

7.2.4 Conceptual Map

The conceptual map provides the interface between the lower levels and the communication system, which is based in natural language. This layer contains the knowledge about the space in the indoor environment together with knowledge about the

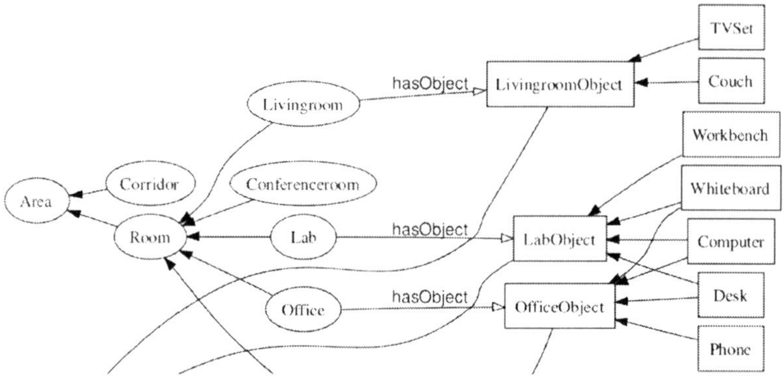

Fig. 7.4 Illustration of a part of the commonsense ontology of an indoor office environment.

entities found therein. In addition, the layer includes the relations between the different components, allowing symbolic reasoning.

Based on the work by Zender [24], the conceptual level is provided with a commonsense OWL ontology [20] of a typical office environment. Part of this ontology is shown in Fig. 7.4. The ontology describes *is-a* relations of room types. Moreover, it shows typical objects found in different places with *has-a* relations. The ontology is created by hand using a priori knowledge about typical office environments. In this ontology, the knowledge about classes and their relations are fixed during the operation of the robot. However, new instances can be added to the representation in real-time by the robot. Using this representation, a reasoner [7] can infer information about the world that is neither given verbally nor actively perceived. In this way, linguistic references to spatial areas can be generated.

The information represented in the ontology is the result of the fusion from *acquired*, *asserted*, and *innate conceptual* knowledge.

7.2.4.1 Acquired Knowledge

The acquired knowledge comprises the information about objects and places that the mobile robot is able to detect in an autonomous manner while moving around. In addition, this knowledge spreads in a bottom-up manner until reaching the conceptual layer. For example, each topological area is represented in the conceptual map as an ontological instance of the type `Area`. As soon as the area is classified as a room or corridor, the instance change its type to a more specific one such as `Room`.

In addition, whenever a new object in the environment is recognized, a new instance of the corresponding type, e.g. `Couch`, is added to the ontology. Moreover, the instance representing the object and the instance of the area where the object is located are related via the *has-a* relation. This process is shown in Fig. 7.2.

7.2.4.2 Asserted Knowledge

The process of acquiring the knowledge about the environment is carried out during a guided tour with the robot [22]. In this tour, a user shows the environment to the robot and names areas and certain objects that should be relevant to it. Typical assertions in a guided tour include "You are in the corridor," or "This is the charging station." These assertions are stored in the conceptual map, either by specifying the type of the current area or by creating a new object instance of the asserted type and linking it to the area instance with the *has-a* relation (cf. Fig. 7.2).

7.2.4.3 Innate Conceptual Knowledge

The innate conceptual knowledge is represented by an ontology that models conceptual commonsense knowledge about an indoor office environment as shown in Fig. 7.4. In the top level we found the two base concepts `Area` and `Object`. `Area` can be further divided into `Room` or `Corridor`. The basic-level subconcepts of `Room` are characterized by the instances of `Object` that are found there with the *has-a* relation. For example, a room with a TV set is represented by the subconcept `LivingRoom` (cf. Fig. 7.2).

7.2.4.4 Inferred Knowledge

Applying a reasoner software [7] to the ontology, a service robot can infer more specific categories for known areas. For example, combining the acquired information that a given topological area is classified as a room and contains a couch, together with the innate conceptual knowledge given in our commonsense ontology, it can be inferred that this area is an instance of `LivingRoom`. On the contrary, if an area is classified as a corridor and the user shows the robot a charging station in that area, the robot can not infer further categories for this area, since according to Fig. 7.2 there are no more subcategories in the original ontology.

7.2.4.5 Ambiguities

The presented conceptual model supports ambiguous classification of areas. That means that the same room can be referred to using different terms. This capability facilitates the interaction with many people simultaneously, since the way people refer to the same room can differ from situation to situation and from speaker to speaker [22]. As an example, a room described as a kitchen by one person can be seen as a recreation room by another person.

7.3 System Integration

The previous approach was implemented in a real robot as part of the explorer scenario in the CoSy project [3]. The robot acquired information about the environment

(a) Couch (b) Bookcase

Fig. 7.5 Two example objects used in the CoSy explorer scenario. (**a**) A couch. (**b**) A bookcase.

using different sensors, namely a laser range finder and a camera. This information was used for object recognition, place classification, and people tracking. All these perception components are also part of the navigation subsystem, which used the sensors for SLAM and motion planning. The complete system is shown in Fig. 7.1.

The approach was implemented and integrated in an ActivMedia PeopleBot mobile platform (robot in Fig. 7.1). The robot was equipped with a SICK laser range finder, which was used for the metric map creation, people following, and for the semantic classification of places. Additionally, a camera was used only for object detection. The detection systems used SIFT features for finding typical objects like a television set, a couch or a bookcase. We recognized instances of objects and not categories [14]. The objects should be previously shown to the robot and learned by it. Examples of objects used for recognition are shown in Fig. 7.5.

The communication with people was done using spoken language only (cf. Fig. 7.1). The user could talk to the robot using a bluetooth headset and the robot replied using a set of speakers mounted on the mobile platform. Voice commands were processed using a real time speech recognition system [17].

The information coming from the sensors was used to create a multi-layered conceptual and spatial representation of the indoor environment. Some of the information needed at the conceptual level to complete this representation was given by the user through spoken dialogues. The system additionally used a linguistic framework that actively supported the map acquisition process and was used for situated dialogue about the environment. The robot could also initiate a clarification dialogue if it detected an inconsistency in its spatial representation, illustrating the mixed-initiative capabilities of the dialogue system [10, 11].

As an additional tool, we used an online viewer for the metric and navigation maps. Examples of snapshots are shown in Fig. 7.6. The output of this program was composed of the lines extracted by our SLAM implementation extended to 3D planes to facilitate the visualization. The viewer showed the different nodes and edges used to construct the navigation map. Nodes corresponding to doorways were drawn bigger and with red color and with an associated doorframe as shown in Fig. 7.6. Finally, the robot and the user were constantly shown in the positions where

(a)

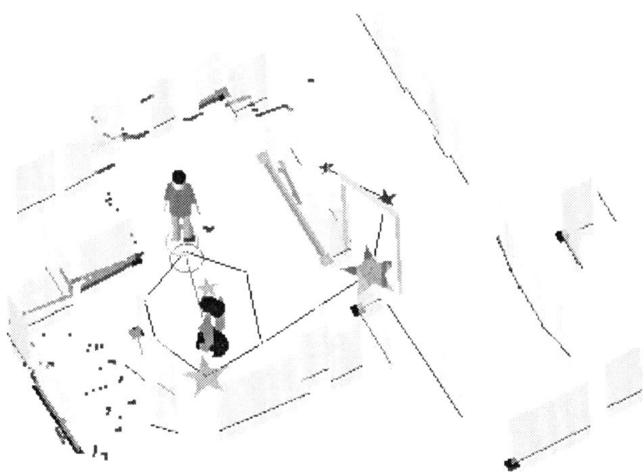

(b)

Fig. 7.6 Two snapshots of the online viewer used during the experiment. The stars indicate the nodes in the navigation map. Small and blue for corridor, small and yellow for room, big and red for doorways and medium and green for the actual position of the robot. Additionally, lines are extended to 3D planes and simulated doorways are drawn for facilitating the visualization. The person is drawn in the position detected by the people following software. (**a**) The robot enters a room after detecting a doorway. (**b**) The complete map of the room is created using lines.

they were localized. The localization of the robot was calculated using the SLAM method introduced in Sect. 7.2.1, while the pose of the person was estimated using people tracking methods based on laser readings only [18].

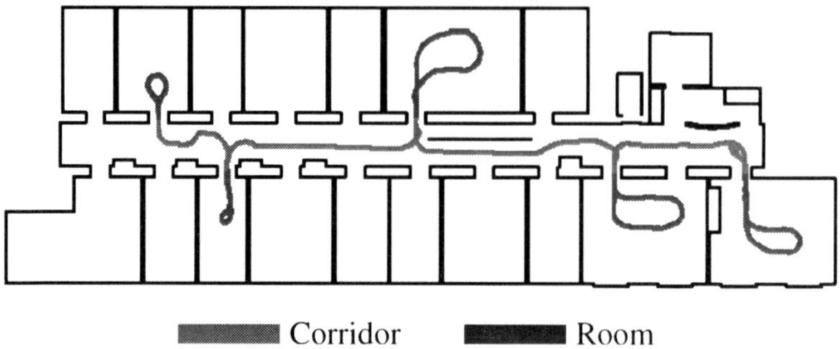

Fig. 7.7 Trajectory followed by the robot to train the classifier for distinguishing between corridor and room. The different places are depicted with distinct colors.

7.4 Demo

In order to show all the functionalities explained in the previous sections, we carried out a demo on the 7th floor of the CAS building at the Royal Institute of Technology in Stockholm. In this demo the robot, together with a user, went through different situations along the environment. The complete demo was carried out non-stop, i.e. we did not stop the robot or restart the system at any moment. The demo was thought of as a test, and for this reason we forced some artificial situations to simulate possible real ones, e.g. a false doorway. The aim of the demo was to show how the robot was able to learn its environment while interacting with a tutor. A video showing the complete demo is available at the CoSy project website [3] under the explorer scenario.

Before running the experiment, some previous knowledge was needed. First, the robot was provided with an ontology representing the general knowledge about the environment. We used the ontology (partially) depicted in Fig. 7.4. Second, the classification of places was based on previous general knowledge about the geometry of rooms and corridors in typical office environments. This knowledge was encoded in a classifier based on laser readings, as explained in Sect. 7.2.2. This classifier was trained using examples of corridors and rooms from real environments, as the one shown in Fig. 7.7. These two kinds of knowledges are independent of the environment used for testing, in the sense that the robot does not need to be physically present in the test environment to acquire the information.

Finally, the robot had to recognize different objects, such as couches or TV sets, using vision (see Fig. 7.5 for some examples). Due to the fact that we performed instance recognition rather than categorization, the objects we wanted to recognize had to be presented to the robot before running the experiment. For this purpose, we positioned the robot in front of these objects, acquired a training image and labeled it with the corresponding term. This term was then added to a small database of objects and also included in the language systems for its posterior use.

The demo started in the corridor, where the robot was positioned close to the charging station. The user activated the robot and told it that it was located at the

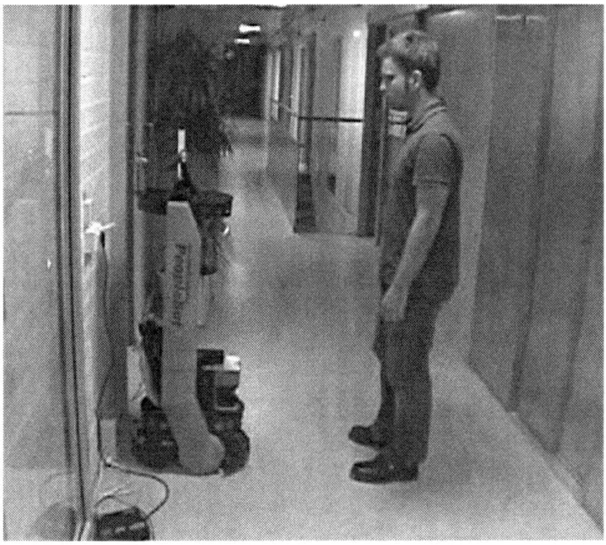

Fig. 7.8 The user wakes up the robot and the demos starts.

charging station. This situation is depicted in Fig. 7.8. The user then asked the robot to follow him. From this point the robot started dropping markers (navigation nodes), which were correctly classified as corridor. Then the person followed by the robot entered a room through a doorway. This doorway was recognized by the robot and the corresponding node was included in the map. From this point, the next nodes were classified as a new area and thus correctly labeled as room.

To show the utility of the clarification dialogues, we forced the robot to detect a false doorway inside a room by putting a bucket close to a table. This created an illusion of a doorway when using only the front laser as sensor. The robot passed through this false doorway and came back to a previously visited node. At this point the robot inferred that there was an inconsistency in the map, and initialized a clarification dialogue asking if there was a door there previously. The user denied this fact and the map was updated accordingly.

The inference of new subconcepts was demonstrated in the following situation. After the false doorway, and while staying inside the room, the robot was asked for the current place and it answered with the indefinite description "a room". The term `Room` was obtained from the classification of the navigation nodes belonging to the current area. A majority vote among these nodes was used in case the node classification was not unanimous. Then the robot was asked to look around. This command activated the vision-based object detection capabilities of the robot. The robot moved and detected a couch, and then a television set. After that, the user asked the robot for the name of the place. Due to the inference over the detected objects and places, the robot categorized the place as a `Livingroom`. Note that previous to the detection of objects the same place was categorized as a `Room`.

Fig. 7.9 Following the order "go to the television", the robot approaches the navigation node from where it saw the television set the last time.

Finally, we showed how the navigation map was used by the robot to come back to previously visited places. After the door opening situation, the robot was asked to go to the television set. The robot then navigated to the node where the television set was last detected (cf. Fig. 7.9). This functionality permitted the user to command the robot to places without the need of giving concrete coordinates. It is also more powerful in the sense that the user may not know the concrete name of the place, but he can remember it as 'the room with a television".

Finally, the robot was commanded to go to the charging station. Again the robot followed the navigation map until it positioned itself on the station, thus finishing the experiment.

7.5 Related Work

Research in spatial representations has yielded different multi-layered environment models. Vasudevan et al. [23] suggest a hierarchical probabilistic representation of space based on objects. The work by Galindo et al. [6] presents an approach containing two parallel hierarchies, spatial and conceptual, connected through anchoring. Inference about places is based on objects found in them. Furthermore, the Hybrid Spatial Semantic Hierarchy (HSSH) was introduced by Beeson et al. [1]. This representation allows a mobile robot to describe the world using different models each containing its own ontology. In contrast to these approaches our implementation uses human augmented mapping to collect information. Furthermore, the communication with the robot is made entirely using natural language and dialogues.

Moreover, our conceptual representation comes from the fusion of acquired, asserted, inferred, and innate knowledge.

There are more cognitively inspired approaches to describe indoor environments. These approaches do not necessarily rely on an exact global self-localization, but rather require the execution of a sequence of local behaviors. Kuipers [12] presented the Spatial Semantic Hierarchy (SSH). Alternatively, the Route Graph model was introduced by Krieg-Brückner et al. [9]. Both theories propose a cognitively inspired multi-layered representation of a map, which is at the same time suitable for robot navigation. Their central layer of abstraction is a topological representation. Our approach differs in that it provides an abstraction layer that can be used for reference resolution of topological entities.

A number of systems have been implemented that permit a robot to interact with humans in their environment. Rhino [2] and Robox [19] are robots that work as tour-guides in museums. Both robots rely on an accurate metric representation of the environment and use limited dialogue to communicate with people. The robot BIRON [21] is endowed with a system that integrates spoken dialogue and visual localization capabilities on a robotic platform. However, these systems differ from ours in the degree to which conceptual spatial knowledge and linguistic meaning are grounded in, and contribute to, situational awareness.

Acquiring a map of the environment with the help of a tutor has been considered in different works. For example, Diosi et al. [4] creates a metric map through a guided tour. The map is then segmented according to the labels given by the instructor. Finally, in the work by Topp et al. [22], a graph based model incorporates information from a user that presents the environment. In contrast to these works, the approach presented in this chapter provides the model with the autonomous acquisition by the robot of knowledge about space and entities.

7.6 Conclusion

In this chapter we presented the semantic classification of places as part of a integrated approach for creating conceptual representations of human-made environments. In this model, the concepts represent spatial and functional properties of typical office indoor environments. This representation is based on multiple maps at different levels of abstraction. The complete system was integrated and tested in a service robot which included a linguistic framework with capabilities for situated dialogue and map acquisition. The presented demo showed that the system was able to provide a high level of human-robot communication and a certain degree of social behavior.

References

1. Beeson, P., MacMahon, M., Modayil, J., Murarka, A., Kuipers, B., Stankiewicz, B.: Integrating multiple representations of spatial knowledge for mapping, navigation, and communication. In: Interaction Challenges for Intelligent Assistants, AAAI Spring Symposium, Stanford, CA, USA (2007)

2. Burgard, W., Cremers, A.B., Fox, D., Hähnel, D., Lakemeyer, G., Schulz, D., Steiner, W., Thrun, S.: Experiences with an interactive museum tour-guide robot. Artificial Intelligence 114(1-2) (2000)
3. CoSy. Cognitive systems for cognitive assistants (2004), `http://www.cognitivesystems.org/`
4. Diosi, A., Taylor, G., Kleeman, L.: Interactive SLAM using laser and advanced sonar. In: Proceedings of the IEEE International Conference on Robotics and Automation, Barcelona, Spain (April 2005)
5. Folkesson, J., Jensfelt, P., Christensen, H.I.: Vision SLAM in the measurement subspace. In: Proceedings of the IEEE International Conference on Robotics and Automation, pp. 30–35 (2005)
6. Galindo, C., Saffiotti, A., Coradeschi, S., Buschka, P., Fernández-Madrigal, J.A., González, J.: Multi-hierarchical semantic maps for mobile robotics. In: Proceedings of the IEEE/RSJ International Conference on Intelligent Robots and Systems, Edmonton, Alberta, Canada (2005)
7. Haarslev, V., Mölle, R.: Racer: A core inference engine for the semantic web. In: Proceedings of the International Workshop on Evaluation of Ontology-based Tools, Florida, USA, October 2003, pp. 27–36 (2003)
8. Hirtle, S.C., Jonides, J.: Evidence for hierarchies in cognitive maps. Memory and Cognition 13, 208–217 (1985)
9. Krieg-Brückner, B., Röfer, T., Carmesin, H.-O., Müller, R.: A taxonomy of spatial knowledge for navigation and its application to the Bremen autonomous wheelchair. In: Freksa, C., Habel, C., Wender, K.F. (eds.) Spatial Cognition 1998. LNCS (LNAI), vol. 1404, pp. 373–397. Springer, Heidelberg (1998)
10. Kruijff, G.-J.M., Zender, H., Jensfelt, P., Christensen, H.I.: Clarification dialogues in human-augmented mapping. In: Proceedings of the 1st ACM Conference on Human-Robot Interaction, Salt Lake City, UT, USA (2006)
11. Kruijff, G.-J.M., Zender, H., Jensfelt, P., Christensen, H.I.: Situated dialogue and spatial organization: What, where...and why? International Journal of Advanced Robotic Systems 4(2) (March 2007)
12. Kuipers, B.: The Spatial Semantic Hierarchy. Artificial Intelligence 119, 191–233 (2000)
13. Latombe, J.C.: Robot Motion Planning. Academic Publishers, Boston (1991)
14. Lowe, D.: Distinctive image features from scale-invariant keypoints. International Journal of Computer Vision 60(2), 91–110 (2004)
15. McNamara, T.P.: Mental representations of spatial relations. Cognitive Psychology 18, 87–121 (1986)
16. Newman, P., Leonard, J., Tardos, J.D., Neira, J.: Explore and return: Experimental validation of real-time concurrent mapping and localization. In: Proceedings of the IEEE International Conference on Robotics and Automation, Washington, D.C., USA, pp. 1802–1809 (2002)
17. Nuance. Nuance speech recognition system developer's manual version 6.2 (1999), `http://www.nuance.com/`
18. Schulz, D., Burgard, W., Fox, D., Cremers, A.B.: People tracking with a mobile robot using sample-based joint probabilistic data association filters. International Journal of Robotics Research 22(2), 99–116 (2003)
19. Siegwart, R., Arras, K.O., Bouabdallah, S., Burnier, D., Froidevaux, G., Greppin, X., Jensen, B., Lorotte, A., Mayor, L., Meisser, M., Philippsen, R., Piguet, R., Ramel, G., Terrien, G., Tomatis, N.: Robox at expo.02: A large scale installation of personal robots. Robotics and Autonomous Systems 42, 203–222 (2003)
20. Smith, M.K., Welty, C., McGuinness, D.L.: OWL web ontology language guide (2004)

21. Spexard, T., Li, S., Wrede, B., Fritsch, J., Sagerer, G., Booij, O., Zivkovic, Z., Terwijn, B., Kröse, B.: BIRON, where are you? - enabling a robot to learn new places in a real home environment by integrating spoken dialog and visual localization. In: Proceedings of the IEEE/RSJ International Conference on Intelligent Robots and Systems (2006)
22. Topp, E.A., Huettenrauch, H., Christensen, H.I., Eklundh, K.S.: Bringing together human and robotic environment representations – a pilot study. In: Proceedings of the IEEE/RSJ International Conference on Intelligent Robots and Systems, Beijing, China (October 2006)
23. Vasudevan, S., Gächter, S., Berger, M., Siegwart, R.: Cognitive maps for mobile robots–an object based approach. In: Proceedings of the IEEE/RSJ IROS 2006 Workshop: From Sensors to Human Spatial Concepts, Beijing, China (2006)
24. Zender, H.: Learning spatial organization through situated dialogue. Master's thesis, Department of Computational Linguistics, Saarland University, Saarbruecken, Germany (2006)

Chapter 8
Semantic Information in Sensor Data*

8.1 Introduction

So far, we have seen how to augment the maps representing environments with semantic information. This additional information was obtained by classifying the laser range data obtained by a mobile robot into some of the classes that represent the different places in the environment.

This chapter deals with semantic information from a different point of view. Instead of classifying the pose of the robot according to the corresponding laser range observation, we classify the observation itself by assigning a semantic label to each of its measurements. The main idea is to classify each laser beam into the class of the entity it hits. In this way, the data provided by the range sensor contains additional semantic information about the environment.

We consider the particular case in which the different range measurements are assigned two possible labels only, namely *person* or *non-person*. We chose these labels because service robots usually operate in populated environments. Therefore, the knowledge about presence and position of people is a key capacity for a service robot. In addition, the information about people will enable the robots to better understand and anticipate intentions and actions.

The application of laser sensors for people detection has been popular because these sensors provide a large field of view and, opposed to vision, are mainly independent from ambient conditions. Typically, the laser scans are located at a height which permits the detection of legs. However, range scans provide little information about legs, because they are represented by short segments composed of few points. Some examples of beams hitting legs of people are shown in Fig. 8.1.

In cluttered environments, like homes or offices, the segments corresponding to people can be easily misclassified with other objects such as tables, chairs and other furniture. Figure 8.2 shows an example scan from a cluttered office environment. While this scan was recorded, several people walked through the office. This example suggests that in cluttered environments, people detection in 2D is difficult even for humans.

---

* This chapter originated from a collaboration with Kai O. Arras.

Ó.M. Mozos: Semantic Labeling of Places with Mobile Robots, STAR 61, pp. 99–108.
springerlink.com

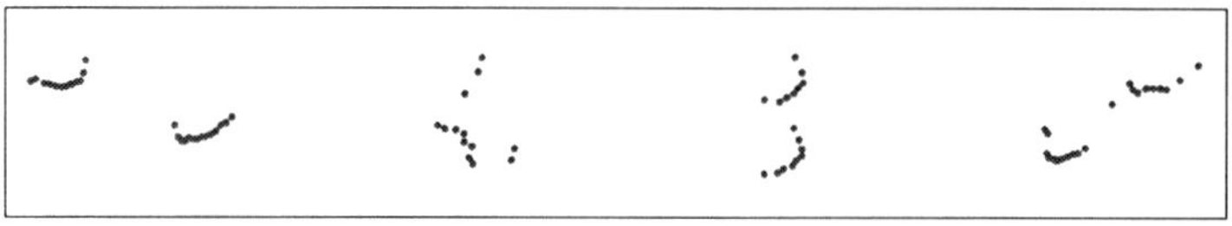

Fig. 8.1 Typical range readings from legs of people. As can be seen, the appearance can change drastically, also because the legs cannot always be separated.

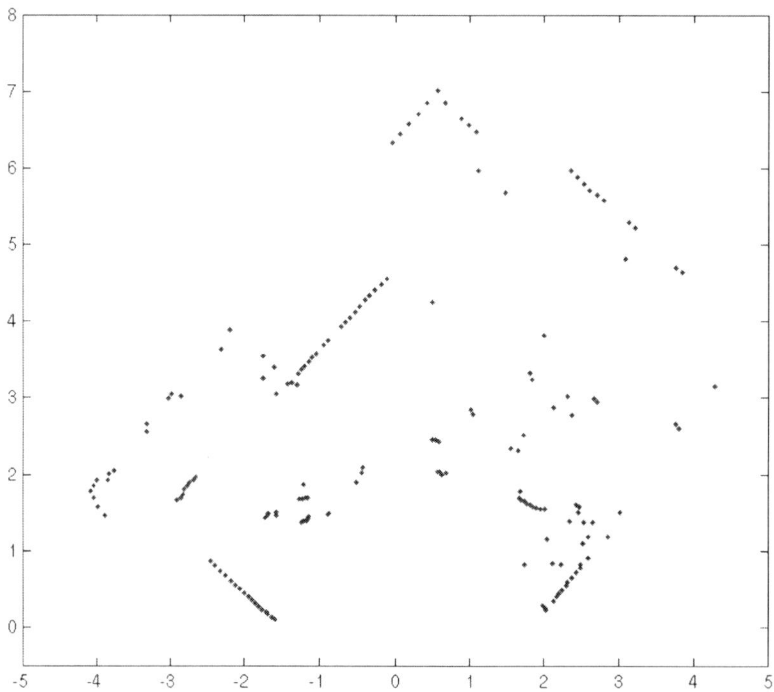

Fig. 8.2 Example scan from a typical office. It seems very difficult to detect which beams are hitting people.

The key idea presented in this chapter is to divide the laser range observations into segments, and then extract several scalar features from them, which encode their geometrical properties. Finally, we apply a supervised learning algorithm based on AdaBoost that uses the scalar features to create a final classifier. This classifier is able to detect segments which are composed of beams hitting legs of people. As a result, each beam in a laser scan is assigned one of the two possible semantic labels *person* or *non-person*.

The rest of the chapter is organized as follows. The next section introduces the segmentation of scans, and the set of geometrical features extracted from each segment. Section 8.3 introduces the algorithm used to detect segments corresponding to people. Experimental results are shown in Sect. 8.4. We discuss related work in Sect. 8.5. Finally, we conclude in Sect. 8.6.

8.2 Feature Extraction

In this section we explain how the geometrical features are extracted from the laser observations. These features will be the input to the learning algorithm.

We assume that the robot is equipped with a range sensor that delivers observations consisting of a set of beams in the form

$$z = \{b_0, ..., b_{M-1}\} .$$

Each beam b_m corresponds to a tuple (ρ_m, d_m), where ρ_m is the angle of the beam relative to the robot and d_m is the length of the beam.

The beams in the observation scan z are split into subsets using a jump distance condition. Whenever two adjacent beams are farther away than a threshold distance, a new subset is initialized. As we will see in the experiments, this simple method results in adequate segmentations for the detection of people. The output of the partitioning procedure is an ordered sequence $\{z_1, ..., z_S\}$ of segments such that $\bigcup z_i = z$. The points of each segment $z_i = \{p_1, p_2, \ldots, p_N\}$ are represented by Cartesian coordinates $p = (p_x, p_y)$, where $p_x = d \cdot \cos(\rho)$ and $p_y = d \cdot \sin(\rho)$, and (ρ, d) are the polar coordinates of the corresponding beam.

Once the laser observation is divided into the different segments, we can proceed to extract the geometrical features from them. We define a feature f as a function that takes a segment z_i as an argument and returns a real value. For each segment z_i we determine the following set of features:

1. Number of points in the segment.
2. The standard deviation of the beam length.
3. The mean average deviation from the median.
4. Jump distance from preceding segment.
5. Jump distance to succeeding segment.
6. The Euclidean distance between the first and last point of a segment.
7. The linearity of the segment.
8. The circularity of the segment.
9. The radius of the circle fitted to the segment.
10. The boundary length of the segment.
11. The boundary regularity of the segment.
12. The average curvature of the segment.
13. The mean angular difference of the segment.
14. Mean speed between two consecutive scans.

Finally, each segment z_i is transformed into a feature vector x in the form

$$x = \tau(z_i) = \{f_1, \ldots, f_L\} .$$

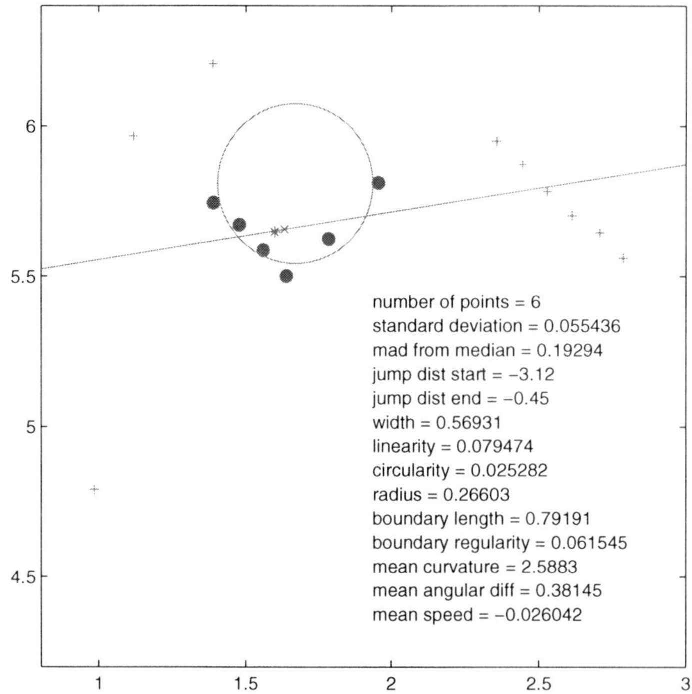

Fig. 8.3 Laser segment with its feature profile. The highlighted points correspond to the segment and the crosses depicts other readings in the scan. The circle and line are fitted to the segment for the linearity and circularity features.

This collection of features constitutes a profile of each segment. An example profile is shown in Fig. 8.3. Since certain features are not defined for less than three points (e.g., circularity, radius) only segments with two or more points are taken into account. Details for the calculation of each feature are given in Appx. C.

8.3 Classification Using Boosting

To classify each of the segments extracted from a laser scan we apply a learning method based on the generalized AdaBoost. As explained in Sect. 2.2.2, the generalized AdaBoost algorithm is a supervised learning algorithm designed to find a binary classifier that discriminates between positive and negative examples. This algorithm boosts the classification performance of a simple learning algorithm by combining a collection of weak classifiers to create a stronger one. The final strong classifier takes the form of a weighted combination of weak classifiers. Large weights are assigned to good classification functions whereas poor functions have small weights.

To classify the different segments of a laser observation, we create a weak classifier for each of the geometrical features f extracted from them. The weak hypotheses have the form

$$h_j(x) = \begin{cases} +1 \text{ if } p_j f_j(x) < p_j \theta_j \\ -1 \text{ otherwise,} \end{cases} \tag{8.1}$$

where θ_j is a threshold and p_j is either -1 or $+1$ and thus representing the direction of the inequality. This form is similar to the one used in Sect. 4.2. Also here the algorithm determines for each weak classifier $h_j(x)$ the optimal values for θ_j and p_j, such that the number of misclassified training examples is minimized. The final learning algorithm used in this chapter was introduced in Sect. 4.2.

The final training set for the generalized AdaBoost algorithm is given by a set of segments represented by their feature vectors together with their corresponding labels

$$\{(x_i, y_i) \mid y_i \in \{+1, -1\}\},$$

where $y_i = +1$ indicates that the beams in segment x_i correspond to a person, and $y_i = -1$ indicates that the beams in segment x_i hit other entities in the environment.

8.4 Experimental Results

The approach presented above has been implemented using a 180° SICK laser range finder. The goal of the experiments is to demonstrate that our simple features can be boosted to a robust classifier for the semantic classification of the beams in a laser scan. Each beam is semantically labeled as *person* or *non-person*.

Throughout the experiments, the sensor was kept stationary and mounted 30 cm above the floor. The corresponding scans were segmented using a jump distance condition with a threshold of 15 cm. For each resulting segment the geometrical features of Sect. 8.2 were calculated.

The complete set of labeled segments was then divided randomly into a training and a test set, each containing approximately 50% of the segments. The training set was employed for learning a strong classifier using the method presented in Sect. 8.3, whereas the test set was used for the evaluations. The segments in both sets were labeled manually with the help of videos recorded during the different experiments.

One important parameter of the AdaBoost algorithm is the number of weak classifiers T used to form the final strong classifier. We need a fast people detector as we want the classifier to work in real time. After several experiments, we found that a value of $T = 100$ weak classifiers was the best trade-off between the error rate and the speed of the classifier.

8.4.1 Corridor and Office Environments

In the first experiment we analyze the performance of our method when used in a corridor. We recorded a total of 540 scans in the corridor of building 79 at the University of Freiburg. This corridor is approximately 20 meters long. The scans were recorded while a person was both moving and standing still (cf. Fig. 8.4(a)). Each scan was divided into segments and for each segment the features #1 to #13

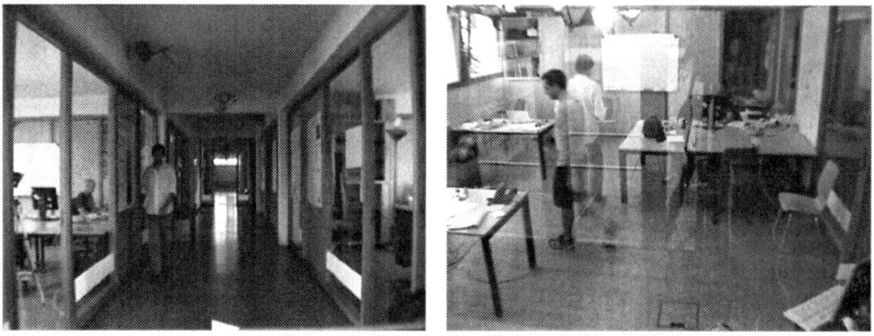

(a) Corridor (b) Office

Fig. 8.4 Images of different environments for the experiments. (**a**) A corridor. (**b**) An office.

were calculated. The total number of segments extracted was 5734. After dividing the segments into a training and a test set, we trained our AdaBoost classifier. The resulting classifications for the test set are shown in Table 8.1. Only 1 from 240 segments (0.42%) corresponding to the person was misclassified (false negatives), whereas 27 from 2616 segments (1.03%) not corresponding to the person were classified as people (false positives).

In a second experiment, we placed the laser in an office that contained tables, chairs, boxes, round shaped trash bins, and other furniture, creating a cluttered environment (cf. Fig. 8.4(b)). An example scan taken in this environment can be shown in Fig. 8.1. In this case two people were in the room during the experiment. As in the previous experiment, the people were moving and occasionally standing still. A total of 791 scans were recorded from which we extracted 13838 segments. The segments were divided into a training and a test set and a strong classifier was learned. Although the office was cluttered with objects and furniture that strongly resemble features of legs, we still obtained an overall classification rate of 97.25%. The confusion matrix is shown in Table 8.2.

In a third experiment we created a common set of segments containing all the segments from both the corridor and the office environment. Again, the set was divided into a training and a test set. Table 8.3 shows the confusion matrix of the resulting classification. Although the error rates slightly increase with respect to Tables 8.1 and 8.2, they still remain under 4%, which in our opinion is a fairly good level. This result demonstrates that a common classifier can be learned using both environments while still obtaining acceptable classification rates.

Table 8.1 Confusion matrix for the corridor environment.

	Detected Label		
True Label	Person	No Person	Total examples
Person	**239** (99.58%)	**1** (0.42%)	**240**
No Person	**27** (1.03%)	**2589** (98.97%)	**2616**

Table 8.2 Confusion matrix for the office environment

	Detected Label		
True Label	Person	No Person	Total examples
Person	**497** (97.45%)	**13** (2.55%)	**510**
No Person	**171** (2.74%)	**6073** (96.26%)	**6244**

Table 8.3 Confusion matrix for the combined corridor and office environment

	Detected Label		
True Label	Person	No Person	Total
Person	**722** (96.27%)	**28** (3.73%)	**750**
No Person	**225** (2.54%)	**8649** (97.46%)	**8874**

8.4.2 Transferring the Classifiers to New Environments

In the following experiment we analyze whether a classifier learned in a particular environment can be used to successfully classify the beam observations in a new environment.

For this purpose we trained our AdaBoost-based classifier using the training set corresponding to the office environment. We then classified the test set from the corridor scenario. Table 8.4 shows the results of this classification. As expected, the number of errors increase compared to the situation in which the training and the test data came from the same environment. However, the classification rates remain above 90%, which indicates that our algorithm yields good generalizations and can also be employed for people detection in new environments.

Table 8.4 Results obtained in the corridor environment using the classifier learned in the office.

	Detected Label		
True Label	Person	No Person	Total
Person	**217** (90.42%)	**23** (9.58%)	**240**
No Person	**112** (4.28%)	**2504** (95.72%)	**2616**

8.4.3 Experiments Including the Motion Feature

In the previous experiments, only the first thirteen geometrical features were used. These features were static and did not take into account changes on the observations during time.

In the experiment of this section, we added the motion feature #14 to the set of features to be fed to the boosting process. All scans from the corridor and the office runs were simultaneously used for training and classification, creating a set similar to the one of Sect. 8.4.1. The results of the classification are shown in Table 8.5. As can be seen, adding the motion feature results only in a marginal improvement

Table 8.5 Classification errors including the motion feature.

	Without Motion Feature	With Motion Feature
False Negatives (%)	3.73	3.47
False Positives (%)	2.54	3.13
Total Error (%)	2.63	3.15

over the classifier without it (cf. Table 8.3). Although the motion feature receives relatively high weight (it is ranked as the third most informative feature), we think that this marginal improvement is simply an expression of the fact that people do not always move.

8.4.4 Best Features for People Detection

As we did in Sect. 3.6.5, we now look into the set of weak classifiers selected by AdaBoost to find the most important. Since each weak classifier represents a feature, this is somehow equivalent to choosing the best set of features. We selected the best weak classifiers according to the importance of their individual weights in the final strong classifier. Table 8.6 lists the five best features for the classifier trained in the corridor, office and combined environments respectively. Note that sometimes the same features occur more than once in a classifier, but they differ in their threshold or weight values.

Analyzing Table 8.6, we can see that the most informative feature in all the environments is the radius of the circle fitted into the segment (feature #9). This feature is an alternative estimation of the size of each segment. The mean angular difference (feature #13) is the second most important feature, quantifying the convexity of the segment. The following features in level of importance are the two jump distances (features #4 and #5). These two features are typically used in the literature for people detection. Finally, we find features #2 and #3, which measure the compactness of the segment. Feature #3 seems to be preferred. The reason for this is likely to be the more robust properties of the mean absolute deviation from the median over the simple standard deviation.

Table 8.6 The best five features for each classifier.

Environment	Five Best Features
Corridor	9, 4, 5, 2, 4
Office	9, 13, 3, 4, 5
Combined	9, 13, 4, 3, 5

8.5 Related Work

In the past, many researchers focused on the problem of tracking people in range scans. One of the most popular approach in this context is to extract legs by detecting moving blobs that appear as local minima in the range image [2, 4, 6, 7]. To this end, two types of features have been quite popular: motion and geometry features. Motion in range data is typically identified by subtracting two subsequent scans. If the robot is moving itself, the scans first have to be aligned, e.g., using scan matching. The drawback of motion features is that only moving people can be found. Topp and Christensen [8] extend the method in [7] with the ability to track also people standing still, which, for instance, is useful for interaction. They report on good results in typical scenarios but also on problems in cluttered environments. They also conclude that either improved motion models or more advanced pattern detection of people is necessary.

Cui et al. [1] pursue a multi-sensor approach to people tracking using multiple laser scanners at foot height and a monocular camera. After registration of the laser data, they extract moving blobs of 15 cm diameter as feet candidates. Two feet candidates at a distance of less than 50 cm are treated as a step candidate.

Geometric features have also been used by Xavier et al. [9]. With a jump distance condition, they split the range image into clusters and apply a set of geometric rules to each cluster to distinguish between lines, circles and legs. A leg is defined as a circle with an additional diameter condition.

In all approaches mentioned above, neither the selection of features nor their thresholds are learned or determined other than by manual design and hand-tuning. This motivates the application of the learning technique presented in this chapter.

Finally, Hähnel et al. [3] have considered the problem of identifying beams in range scans that are reflected by dynamic objects. They consider the individual beams independently and apply EM to determine whether or not a beam has been reflected by a dynamic object such as a person. Our method, in contrast, considers groups of beams and classifies the entire groups according to their properties.

Parts of the approach presented in this chapter have been used in posterior works on people detection and/or tracking. Zivkovic and Kröse [10] apply our method to detect people in 2D laser range data. The detection of people is also done using vision. Both modalities are combined to create a robust people detector. Finally, the work by Premebida et al. [5] use some of the geometrical features presented in this chapter for the detection and tracking of objects in laser readings.

8.6 Conclusion

This chapter addressed the problem of adding semantic information about people in sensor readings. Our approach applies the AdaBoost algorithm to learn a robust classifier from simple features, and it identifies groups of beams that correspond to legs of people. The method has been implemented and applied in cluttered office environments. In practical experiments carried out in different environments we obtained encouraging detection rates.

From the features selected by AdaBoost we can conclude that the shape of people in range data is best recognized by a radius feature, a convexity feature, a local minimum feature and a robust compactness feature.

Although in this chapter we concentrated on the detection of people, we think that the approach presented here can be easily extended to add semantic information from other objects in the environment to the beams in a laser scan.

References

1. Cui, J., Zha, H., Zhao, H., Shibasaki, R.: Tracking multiple people using laser and vision. In: Proceedings of the IEEE/RSJ International Conference on Intelligent Robots and Systems, Alberta, Canada (2005)
2. Fod, A., Howard, A., Mataric, M.J.: Laser-based people tracking. In: Proceedings of the IEEE International Conference on Robotics and Automation (2002)
3. Hähnel, D., Triebel, R., Burgard, W., Thrun, S.: Map building with mobile robots in dynamic environments. In: Proceedings of the IEEE International Conference on Robotics and Automation (2003)
4. Kleinhagenbrock, M., Lang, S., Fritsch, J., Lömker, F., Fink, G.A., Sagerer, G.: Person tracking with a mobile robot based on multi-modal anchoring. In: Proceedings of the IEEE International Workshop on Robot and Human Interactive Communication, Berlin, Germany (2002)
5. Premebida, C., Monteiro, G., Nunes, U., Peixoto, P.: Lidar and vision-based approach for pedestrian and vehicle detection and tracking. In: Proceedings of the IEEE Intelligent Transportation Systems Conference, Seattle,Washington, USA, pp. 1044–1049 (2007)
6. Scheutz, M., McRaven, J., Cserey, G.: Fast, reliable, adaptive, bimodal people tracking for indoor environments. In: Proceedings of the IEEE/RSJ International Conference on Intelligent Robots and Systems, Sendai, Japan (2004)
7. Schulz, D., Burgard, W., Fox, D., Cremers, A.B.: People tracking with a mobile robot using sample-based joint probabilistic data association filters. International Journal of Robotics Research 22(2), 99–116 (2003)
8. Topp, E.A., Christensen, H.I.: Tracking for following and passing persons. In: Proceedings of the IEEE/RSJ International Conference on Intelligent Robots and Systems, Alberta, Canada (2005)
9. Xavier, J., Pacheco, M., Castro, D., Ruano, A., Nunes, U.: Fast line, arc/circle and leg detection from laser scan data in a player driver. In: Proceedings of the IEEE International Conference on Robotics and Automation (2005)
10. Zivkovic, Z., Kröse, B.: Part based people detection using 2D range data and images. In: Proceedings of the IEEE/RSJ International Conference on Intelligent Robots and Systems, San Diego, CA, USA, pp. 214–219 (2007)

Chapter 9
Conclusion

This book presented different approaches for adding semantic information to the representations of indoor environments. We concentrated on extending the information in the maps created by a mobile robot with labels that represent different places in the environment. These places have different functionalities, such as corridors, offices or kitchens. Moreover, throughout this book we have seen how the semantic information about places can improve the capabilities of mobile robots in different domains including human-robot interaction, localization, and exploration.

We first presented a technique based on supervised learning that enabled a mobile robot to recognize the different places in an indoor environment using a laser sensor. To carry out this classification the robot should first take observations and then extract some features from them. These features were later used to recognize the different places. As main observations we used the range measurements of laser range finders, from which several features were extracted that encoded their geometrical properties.

The learning method used for classifying the different places was based on the AdaBoost algorithm. The input to this algorithm was the features extracted from the observations, and as output we obtained a strong classifier which included the more informative features for each place.

The geometrical features are quite good candidates for generalization, since they encode space information. We saw in Chapts. 3 to 5 that the strong classifier created with geometrical features could successfully be transferred among different environments. The main reason is that indoor environments usually contain the same type of places, as for instance, corridors, doorways and rooms. These places share similar structures between the different indoor environments: corridors are typically elongated, and rooms are usually more compact and cluttered. These common characteristics permitted the robot to learn a classifier in one environment and recognized similar places in different ones.

In addition, we used vision sensors to increase the number of places to classify. The main problem with vision observations was to select the features that maintained a good generalization in the classifier over different places. We opted for counting the number of specific objects that appeared in a panoramic image

Ó.M. Mozos: Semantic Labeling of Places with Mobile Robots, STAR 61, pp. 109–111.
springerlink.com

taken by the robot. The selection of these features was motivated by the fact that typical objects appear at different places with different probabilities. For example, the probability of finding a computer monitor in an office is larger than finding one in a kitchen. Again these features are usually very common in several indoor environments.

The previous approach for semantic classification was used to classify a single pose of a mobile robot using the laser and image features. However, this method did not take into account the classification of neighboring poses when labeling the current one. To include this information, we extended the approach with some probabilistic techniques. We first smoothed the classification of all poses in an environment using probabilistic relaxation, and alternatively an instance-based associative Markov network. Both approaches improved the final classification using neighboring information, and allowed the robot to extract compact regions of the environment and create a topological map.

Mobile robots are dynamic agents that move along different trajectories. When operating in indoor environments, the robots usually have a moderate velocity and a relatively continuous movement. Furthermore, certain transitions between places in a trajectory are more likely. For example, to go from the kitchen to the office a robot should traverse a doorway first. This transitional information was encoded in a hidden Markov model and successfully applied to smooth the classification of the poses of a mobile robot along a trajectory. Different results using this approach were presented in Chapt. 5.

The semantic information about places can improve other typical robotics tasks. The main idea is that mobile robots can use the intrinsic information of human-made environments to improve their actions. In particular, we showed how the information about places could improve the performance of a team of mobile robots during exploration. The results of the experiments in Chapt. 6 demonstrated that places such as corridors are better exploration targets as they lead to other rooms.

Another typical problem is the localization of mobile robots. In this problem a robot must determine its pose relative to a given map. Recognizing the type of place in which the robot is located can be seen as a high level localization. If the robot is situated in an office, then other places can be discarded, and the robot can concentrate on hypotheses that belong only to office places. This idea was presented in Chapt. 6 together with experiments that corroborated its usefulness.

Since one of the main goals of the semantic labeling is to share spatial terms with humans, it seems necessary to develop robotic systems that can communicate these concepts to people. In Chapt. 7, we introduced an integrated system for conceptual representations of indoor environments. This system included a linguistic framework with capabilities for situated dialogue and map acquisition. The experiments of this chapter showed the interaction capabilities of the system, and demonstrated how a high level conceptual representation could be created based on language communication and semantic information about places.

The semantic information can also represent other kind of objects in the environment. In Chapt. 8, we presented an approach to include the semantic information directly in the data beams from a laser range finder. In this way, much richer information was available from the sensor.

In summary, this book presented many innovative approaches in respect to semantic information about places using mobile robots. As we described in the related literature of several chapters, various posterior works have applied and extended some of the ideas presented here. This indicates that a lot of work can still be done.

In future work it would be interesting to move from the supervised approach presented in this book to other methods with less supervision. One possibility could be to use semi-supervised techniques, in which the robot autonomously creates an initial classification of the environment. This classification can be corrected later on by the user. Another option could be to leave the robot to create a totally unsupervised classification of the places it visits.

In any case, it seems that information coming from the user is important, since someone has to decide how to name the different places. This last issue bring us to the problem of personalization: people can describe the same place with different terms. For example, what to one person might be a living room, could be a sitting room to another. It could be interesting to study approaches able to cope with this flexibility. Some solutions were already presented in Chapt. 7.

Finally, we think that the semantic labeling of places is a research area which will have a high impact in the future of mobile robotics.

Appendix A
Simple Features for Place Classification Using Raw Laser Beams

In this appendix we introduce the mathematical formulation for the first set of simple geometrical features used for place classification. We assume that the mobile robot is equipped with a 360° field of view range sensor. Each observation z contains a set of beams in the form

$$z = \{b_0, \ldots, b_{M-1}\} .$$

Each beam b_i consists of a tuple (ρ_m, d_m) where ρ_m is the angle of the beam relative to the robot and d_m is the length of the beam. We define a feature f as a function that takes as argument one observation and returns a real value: $f : \mathrm{Z} \rightarrow \mathbb{R}$, where Z is the set of all possible observations.

The set of features presented here is calculated from the raw laser beams in the observed scan z, and corresponds to the set A introduced in Sect. 3.4 together with its extension from Sect. 4.4. All features are rotational invariant to make the classification of a pose dependent only on the (x, y)-position of the robot and not on its orientation. Most of the features are standard geometrical features often used in shape analysis [1, 2, 3, 4, 5].

A.1 The Average Difference between the Length of Two Consecutive Beams

The average difference between the length of two consecutive beams f_{average} is defined as

$$f_{\text{average}} = \frac{1}{M} \sum_{i=0}^{M-1} \left| d_i - d_{[(i+1) \bmod M]} \right| , \tag{A.1}$$

where mod indicates the module function.

A.2 The Standard Deviation of the Difference between the Length of Two Consecutive Beams

The standard deviation of the difference between the length of two consecutive beams f_{std} is defined as

$$f_{\text{std}} = \sqrt{\frac{1}{M} \sum_{i=0}^{M-1} \left(\left| d_i - d_{[(i+1) \bmod M]} \right| - f_{\text{average}} \right)^2}, \tag{A.2}$$

where f_{average} is the feature defined in (A.1).

A.3 The Average Difference between the Length of Two Consecutive Beams Considering Max-Range Values

The value max-range is a threshold θ indicating the maximum length d_i of a beam. Using this threshold, we define the function $\text{length}_\theta(b_i)$ as

$$\text{length}_\theta(b_i) = \begin{cases} d_i \text{ if } d_i \leq \theta \\ \theta \text{ otherwise}, \end{cases} \tag{A.3}$$

The feature representing the average difference between the length of two consecutive beams using max-range $f_{\text{average},\theta}$ is then defined as

$$f_{\text{average},\theta} = \frac{1}{M} \sum_{i=0}^{M-1} \left| \text{length}_\theta(b_i) - \text{length}_\theta(b_{[(i+1) \bmod M]}) \right| . \tag{A.4}$$

A.4 The Standard Deviation of the Difference between the Length of Two Consecutive Beams Considering Max-Range Values

The standard deviation of the difference between the length of two consecutive using max-range $f_{\text{std},\theta}$ is defined as

$$f_{\text{std},\theta} = \sqrt{\frac{1}{M} \sum_{i=0}^{M-1} \left(\left| \text{length}_\theta(b_i) - \text{length}_\theta(b_{[(i+1) \bmod M]}) \right| - f_{average,\theta} \right)^2}, \tag{A.5}$$

where $f_{\text{average},\theta}$ is the feature defined in (A.4), and $\text{length}_\theta(b_i)$ is the function defined in (A.3).

A.5 The Average Beam Length

The average beam length $f_{\overline{d}}$ is defined as

$$f_{\overline{d}} = \frac{1}{M} \sum_{i=0}^{M-1} d_i \,. \tag{A.6}$$

A.6 The Standard Deviation of the Beam Length

The standard deviation f_σ of the beam length is defined as

$$f_\sigma = \sqrt{\frac{1}{M} \sum_{i=0}^{M-1} \left(d_i - f_{\overline{d}}\right)^2}\,, \tag{A.7}$$

where $f_{\overline{d}}$ is the feature defined in (A.6).

A.7 Number of Gaps

Two consecutive beams build a gap if the difference between their lengths is greater than a given threshold θ. An example of a gap is shown in Fig. A.1. Formally, we define a gap with threshold θ as

$$\mathrm{gap}_\theta(b_i, b_j) = \begin{cases} 1 & \text{if } \left|d_i - d_j\right| > \theta \\ 0 & \text{otherwise}\,. \end{cases} \tag{A.8}$$

The feature $f_{\mathrm{gaps}.\theta}$, representing the total number of gaps in a laser scan, is defined as

$$f_{\mathrm{gaps}.\theta} = \sum_{i=0}^{M-1} \mathrm{gap}_\theta\left(b_i, b_{[(i+1) \bmod M]}\right). \tag{A.9}$$

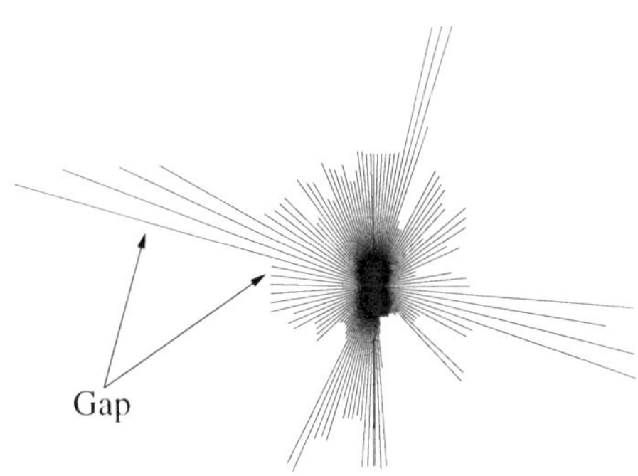

Fig. A.1 Example of a gap in a laser scan.

A.8 The Number of Beams Lying on Lines Extracted from the Range Scan

The number of beams lying on lines extracted from the range scan is calculated using the method by Sack and Burgard [6].

A.9 The Euclidean Distance between the Two Points Corresponding to Two Global Minima

The Euclidean distance between the two points corresponding to two consecutive global minima was designed to help in the classification of doors.

As an example, Fig. A.2(a) shows a scan taken while the robot was situated inside a doorframe. Plotting the length of the beams of this observation, we obtain a graph like the one shown in Fig. A.2(b). We look in this graph for two global minima $p_1 = (x_1, y_1)$ and $p_2 = (x_2, y_2)$, and the Euclidean distance between these two points is then calculated as

$$f_{\mathrm{d-minima}} = \sqrt{(x_1 - x_2)^2 + (y_1 - y_2)^2}\,. \tag{A.10}$$

A.10 The Angular Distance between the Two Points Corresponding to Two Global Minima

The two points (p_1, p_2) corresponding to the global minima calculated in Sect. A.9 correspond to the end of two beams, i.e., b_1 and b_2. The angular distance between these two beams is used as a feature, in the form:

$$f_{\rho-\mathrm{minima}} = |\rho_1 - \rho_2|\,, \tag{A.11}$$

where ρ_i is the angle of beam b_i relative to the robot.

A.11 The Average of the Relation between the Lengths of Two Consecutive Beams

The average relation between the lengths of two consecutive beams $f_{\mathrm{average-rel}}$ is defined as

$$f_{\mathrm{average-rel}} = \frac{1}{M} \sum_{i=0}^{M-1} \frac{d_i}{d_{[(i+1) \bmod M]}}\,. \tag{A.12}$$

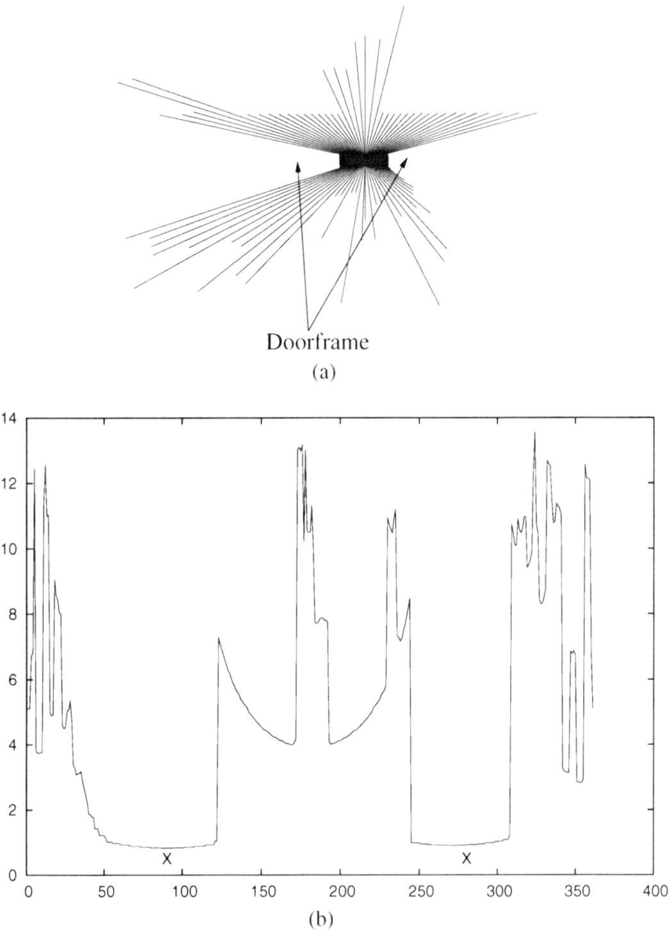

Fig. A.2 Laser scan collected when the robot was situated inside a doorframe. The arrows in (**a**) indicate the two minima which could indicate a doorframe. The plot in (**b**) represents the length of the beams. The two marks **X** indicate the two global minima from (**a**).

A.12 The Standard Deviation of the Relation between the Length of Two Consecutive Beams

The standard deviation of the relation between the length of consecutive beams $f_{\text{std-rel}}$ is defined as

$$f_{\text{std-rel}} = \sqrt{\frac{1}{M}\sum_{i=0}^{M-1}\left(\frac{d_i}{d_{[(i+1) \bmod M]}} - f_{\text{average-rel}}\right)^2}, \qquad \text{(A.13)}$$

where $f_{\text{average-rel}}$ is the feature defined in (A.12).

A.13 The Average of the Normalized Beam Length

The average of the normalized beam length $f_{\text{average-norm}}$ is defined as

$$f_{\text{average-norm}} = \frac{1}{M} \sum_{i=0}^{M-1} \frac{d_i}{d_{max}}, \tag{A.14}$$

where d_{max} corresponds to

$$d_{max} = \max d_i \ \forall i . \tag{A.15}$$

A.14 The Standard Deviation of the Normalized Beam Length

The standard deviation of the normalized beam length $f_{\text{std-norm}}$ is defined as

$$f_{\text{std-norm}} = \sqrt{\frac{1}{M} \sum_{i=0}^{M-1} \left(\frac{d_i}{d_{max}} - f_{\text{average-norm}} \right)^2}, \tag{A.16}$$

where $f_{\text{average-norm}}$ corresponds to the feature defined in (A.14).

A.15 The Number of Relative Gaps

Two consecutive beams build a relative gap if the relation between their lengths is greater than a given threshold θ. Formally, we define a relative gap with threshold θ as

$$\text{rgap}_\theta(b_i, b_j) = \begin{cases} 1 \text{ if } \frac{d_i}{d_j} > \theta \\ 0 \text{ otherwise}. \end{cases} \tag{A.17}$$

The feature $f_{\text{rgaps},\theta}$, which represents the total number of relative gaps in a scan, is defined as

$$f_{\text{rgaps},\theta} = \sum_{i=0}^{M-1} \text{rgap}_\theta(b_i, b_{[(i+1) \bmod M]}) . \tag{A.18}$$

A.16 Kurtosis

We define the kurtosis f_{kurtosis} of a scan as

$$f_{\text{kurtosis}} = \frac{\sum_{i=0}^{M-1} \left(d_i - f_{\overline{d}}\right)^4}{M \cdot f_\sigma^4} - 3, \tag{A.19}$$

where $f_{\overline{d}}$ and f_σ are the features defined in (A.6) and (A.7) respectively.

References

1. Gonzalez, R.C., Wintz, P.A.: Digital Image Processing. Addison-Wesley Publishing Inc., Reading (1987)
2. Haralick, R.M., Shapiro, L.G.: Computer and Robot Vision. Addison-Wesley Publishing Inc., Reading (1992)
3. Loncaric, S.: A survey of shape analysis techniques. Pattern Recognition 31(8), 983–1001 (1998)
4. O'Rourke, J.: Computational Geometry in C, 2nd edn. Cambridge University Press, Cambridge (1998)
5. Russ, J.C.: The Image Processing Handbook. CRC Press, Boca Raton (1992)
6. Sack, D., Burgard, W.: A comparison of methods for line extraction from range data. In: Proceedings of the IFAC Symposium on Intelligent Autonomous Vehicles, Lisabon, Portugal (2004)

Appendix B
Simple Features for Place Classification Using a Polygonal Approximation $\mathrm{Pol}(z)$ of the Scan

In this appendix we introduce the mathematical formulation for the second set of simple geometrical features used for place classification. We assume that the mobile robot is equipped with a 360° field of view range sensor. Each observation z contains a set of beams in the form

$$z = \{b_0, ..., b_{M-1}\} .$$

Each beam b_i consists of a tuple (ρ_m, d_m) where ρ_m is the angle of the beam relative to the robot and d_m is the length of the beam. A polygonal approximation $\mathrm{Pol}(z)$ of the area covered by each observation is calculated. The vertices v_i of this closed polygon $\mathrm{Pol}(z)$ correspond to the coordinates of the end-points of each beam b_i of z relative to the robot in the form

$$\mathrm{Pol}(z) = \{(d_m \cos\rho_m, d_m \sin\rho_m) \mid m = 0, \ldots, M-1\} . \tag{B.1}$$

We define a feature f as a function that takes as argument one observation and returns a real value: $f : \mathrm{P} \rightarrow \mathbb{R}$, where P is the set of all possible polygon approximations.

The set of features presented here is calculated from the polygonal approximation of the observed scan z, and corresponds to the set B introduced in Sect. 3.4 together with its extension from Sect. 4.4. All features are rotational invariant to make the classification of a pose dependent only on the (x, y)-position of the robot and not of its orientation. Most of the features are standard geometrical features often used in shape analysis [1, 2, 3, 4, 5].

B.1 Area of $\mathrm{Pol}(z)$

The area of the polygonal approximation $\mathrm{Pol}(z)$ is given by

$$f_{\mathrm{Area}} = \frac{1}{2} \sum_{i=0}^{M-1} (x_i y_{[(i+1) \bmod M]} - x_{[(i+1) \bmod M]} y_i) . \tag{B.2}$$

B.2 Perimeter of Pol(z)

The perimeter of the polygonal approximation Pol(z) is given by

$$f_{\text{Perimeter}} = \sum_{i=0}^{M-1} \operatorname{dist}(v_i, v_{[(i+1) \bmod M]}), \tag{B.3}$$

where

$$\operatorname{dist}(v_i, v_{i+1}) = \sqrt{(x_i - x_{[(i+1) \bmod M]})^2 + (y_i - y_{[(i+1) \bmod M]})^2}. \tag{B.4}$$

B.3 The Area of Pol(z) Divided by Its Perimeter

The area of Pol(z) divided by its perimeter f_{AP} is defined as

$$f_{AP} = \frac{f_{\text{Area}}}{f_{\text{Perimeter}}}, \tag{B.5}$$

where f_{Area} and $f_{\text{Perimeter}}$ correspond to features (B.2) and (B.3) respectively.

B.4 The Mean Distance between the Centroid and the Shape Boundary of Pol(z)

The centroid $c = (c_x, c_y)$ of Pol(z) is defined as

$$c_x = \frac{1}{6 \cdot f_{\text{Area}}} \sum_{i=0}^{M-1} (x_i + x_{[(i+1) \bmod M]})(x_i y_{[(i+1) \bmod M]} - x_{[(i+1) \bmod M]} y_i), \tag{B.6}$$

$$c_y = \frac{1}{6 \cdot f_{\text{Area}}} \sum_{i=0}^{M-1} (y_i + y_{[(i+1) \bmod M]})(x_i y_{[(i+1) \bmod M]} - x_{[(i+1) \bmod M]} y_i), \tag{B.7}$$

where f_{Area} is the feature defined in (B.2). The mean distance between the centroid and the shape boundary of Pol(z) is thus calculated as

$$f_{\text{mean-shape}} = \frac{1}{M} \sum_{i=0}^{M-1} \operatorname{dist}(v_i, c), \tag{B.8}$$

where

$$\operatorname{dist}(v_i, c) = \sqrt{(x_i - c_x)^2 + (y_i - c_y)^2} \tag{B.9}$$

B.5 The Standard Deviation of the Distances between the Centroid and the Shape Boundary of Pol(z)

The standard deviation of the distances between the centroid and the shape boundary of Pol(z) is given by

$$f_{\text{std-shape}} = \sqrt{\frac{1}{M}\sum_{i=0}^{M-1}\left(\text{dist}(v_i,c) - f_{\text{mean-shape}}\right)^2}, \tag{B.10}$$

where $f_{\text{mean-shape}}$ is defined in (B.8), and the function $\text{dist}(v_i,c)$ is defined in (B.9).

B.6 Similarity Invariant Descriptors Based on the Fourier Transformation of Pol(z)

To calculate the Fourier coefficients of $\text{Pol}(z)$ we transform each vertex $v_i \in \mathbb{R}^2$ into a complex number $\tilde{v}_i \in \mathbb{C}$ in the form

$$v_i = (x_i, y_i), \Longrightarrow \tilde{v}_i = x_i + y_i j, \quad j = \sqrt{-1}. \tag{B.11}$$

The Fourier coefficients $\{c_{-n}, \ldots, c_{-1}, c_0, c_1, \ldots, c_n\}$ of $\text{Pol}(z)$ are calculated as

$$c_0 = \frac{1}{2T}\sum_{k=0}^{M-1}\left(\tilde{v}_k + \tilde{v}_{[(k+1) \bmod M]}\right) \mid \triangle\tilde{v}_k \mid \tag{B.12}$$

$$c_n = \frac{T}{(2\pi n)^2}\sum_{k=0}^{M-1}\left(\triangle s_{[(k+1) \bmod M]} - \triangle s_k\right)e^{-j\pi\left(\frac{2\pi}{T}\right)t_k}, \tag{B.13}$$

with

$$\triangle\tilde{v}_i = \tilde{v}_{[(i+1) \bmod M]} - \tilde{v}_i \tag{B.14}$$

$$\triangle s_i = \triangle\tilde{v}_i / |\triangle\tilde{v}_i| \tag{B.15}$$

$$t_k = \sum_{i=0}^{k-1}|\triangle\tilde{v}_i| \quad k > 0, t_0 = 0 \tag{B.16}$$

$$T = \text{perimeter of } P(z). \tag{B.17}$$

The Fourier descriptors $\{\tilde{x}_{-n}, \ldots, \tilde{x}_0, \ldots, \tilde{x}_n\}$, which are invariant to similarity, that is, translation, rotation and scale, are calculated as

$$\left\{\tilde{x}_n := \frac{|c_n|}{|c_1|}e^{j(\Phi_n + (1-n)\Phi_2 - (2-n)\Phi_1)}\right\} \tag{B.18}$$

$$\Phi_n = \text{phase of } c_n. \tag{B.19}$$

B.7 Major Axis of the Ellipse That Approximates Pol(z)

Using the first two Fourier coefficients (c_{-1}, c_1) of $\text{Pol}(z)$ (cf. Sect. B.6), we calculate the major axis of an ellipse that approximates the polygon $\text{Pol}(z)$ as

$$f_{\text{Ma}} = |c_1| + |c_{-1}|. \tag{B.20}$$

B.8 Minor Axis of the Ellipse That Approximates Pol(z)

Using the first two Fourier coefficients (c_{-1}, c_1) of Pol(z) (cf. Sect. B.6), we calculate the minor axis of an ellipse that approximates the polygon Pol(z) as

$$f_{\mathrm{Mi}} = ||c_1| - |c_{-1}||\,. \tag{B.21}$$

B.9 Invariant Moments of Pol(z)

The central moments μ_{pq} of Pol(z) up to three are defined as

$$\mu_{10} = \sum_{i=0}^{M-1}\sum_{j=0}^{M-1} x_i - \bar{x}^1 (y_j - \bar{y})^0\,, \tag{B.22}$$

$$\mu_{01} = \sum_{i=0}^{M-1}\sum_{j=0}^{M-1} (x_i - \bar{x})^0 (y_j - \bar{y})^1\,, \tag{B.23}$$

$$\mu_{11} = \sum_{i=0}^{M-1}\sum_{j=0}^{M-1} (x_i - \bar{x})^1 (y_j - \bar{y})^1\,, \tag{B.24}$$

$$\mu_{20} = \sum_{i=0}^{M-1}\sum_{j=0}^{M-1} (x_i - \bar{x})^2 (y_j - \bar{y})^0\,, \tag{B.25}$$

$$\mu_{02} = \sum_{i=0}^{M-1}\sum_{j=0}^{M-1} (x_i - \bar{x})^0 (y_j - \bar{y})^2\,, \tag{B.26}$$

$$\mu_{30} = \sum_{i=0}^{M-1}\sum_{j=0}^{M-1} (x_i - \bar{x})^3 (y_j - \bar{y})^0\,, \tag{B.27}$$

$$\mu_{03} = \sum_{i=0}^{M-1}\sum_{j=0}^{M-1} (x_i - \bar{x})^0 (y_j - \bar{y})^3\,, \tag{B.28}$$

$$\mu_{12} = \sum_{i=0}^{M-1}\sum_{j=0}^{M-1} (x_i - \bar{x})^1 (y_j - \bar{y})^2\,, \tag{B.29}$$

$$\mu_{21} = \sum_{i=0}^{M-1}\sum_{j=0}^{M-1} (x_i - \bar{x})^2 (y_j - \bar{y})^1\,, \tag{B.30}$$

with

$$\bar{x} = \frac{1}{M}\sum_{i=0}^{M-1} x_i\,, \tag{B.31}$$

and

$$\bar{y} = \frac{1}{M} \sum_{j=0}^{M-1} y_j , \tag{B.32}$$

The normalized central moments of Pol(z), denoted η_{pq}, are defined as

$$\eta_{pq} = \frac{\mu_{pq}}{\mu_{00}^{\gamma}} , \tag{B.33}$$

where

$$\gamma = \frac{p+q}{2} + 1 , \tag{B.34}$$

for $p+q = 2, 3, \ldots$

A set of seven invariant moments with respect translation, rotation and scale can be derived from the second and third normalized central moments as

$$\phi_1 = \eta_{20} + \eta_{02} , \tag{B.35}$$

$$\phi_2 = (\eta_{20} - \eta_{02})^2 + 4\eta_{11}^2 , \tag{B.36}$$

$$\phi_3 = (\eta_{30} - 3\eta_{12})^2 + (3\eta_{21} - \eta_{03})^2 , \tag{B.37}$$

$$\phi_4 = (\eta_{30} + \eta_{12})^2 + (\eta_{21} + \eta_{03})^2 , \tag{B.38}$$

$$\begin{aligned} \phi_5 = {} & (\eta_{30} - 3\eta_{12})(\eta_{30} + \eta_{12}) \left[(\eta_{30} + \eta_{12})^2 - 3(\eta_{21} + \eta_{03})^2\right] \\ & + (3\eta_{21} - \eta_{03})(\eta_{21} + \eta_{03}) \left[3(\eta_{30} - \eta_{12})^2 - (\eta_{21} + \eta_{03})^2\right] , \end{aligned} \tag{B.39}$$

$$\begin{aligned} \phi_6 = {} & (\eta_{20} - \eta_{02}) \left[(\eta_{30} + \eta_{12})^2 - (\eta_{21} + \eta_{03})^2\right] \\ & + 4\eta_{11}(\eta_{30} + \eta_{12})(\eta_{21} + \eta_{03}) , \end{aligned} \tag{B.40}$$

$$\begin{aligned} \phi_7 = {} & (3\eta_{21} - \eta_{03})(\eta_{30} - \eta_{12}) \left[(\eta_{30} - \eta_{12})^2 - 3(\eta_{21} + \eta_{03})^2\right] \\ & + (3\eta_{12} - \eta_{30})(\eta_{21} + \eta_{03}) \left[3(\eta_{30} + \eta_{12})^2 - (\eta_{21} + \eta_{03})^2\right] . \end{aligned} \tag{B.41}$$

B.10 The Normalized Feature of Compactness of Pol(z)

The normalized feature of compactness f_{cmp} of Pol(z) is calculated as

$$f_{\text{cmp}} = \frac{f_{\text{Area}}}{\mu_{20} + \mu_{02}}, \tag{B.42}$$

where f_{Area} is the feature defined in (B.2), and μ_{20} and μ_{02} are the central moments of second order calculated in Sect B.9.

B.11 The Normalized Feature of Eccentricity of $\text{Pol}(z)$

The normalized feature of eccentricity f_{ect} of $\text{Pol}(z)$ is defined as

$$f_{\text{ect}} = \frac{\sqrt{(\mu_{20} + \mu_{02})^2 + 4\mu_{11}^2}}{\mu_{20} + \mu_{02}}, \tag{B.43}$$

where $\mu_{20}, \mu_{02}, \mu_{11}$ are the central moments of second order calculated in Sect. B.9.

B.12 The Form Factor of $\text{Pol}(z)$

The form factor of $\text{Pol}(z)$ is given by

$$f_{\text{f-factor}} = \frac{4\pi f_{\text{Area}}}{\sqrt{f_{\text{Perimeter}}}}, \tag{B.44}$$

where f_{Area} and $f_{\text{Perimeter}}$ are the features defined in (B.2) and (B.3) respectively.

B.13 Circularity of $\text{Pol}(z)$

The circularity of $\text{Pol}(z)$ is defined as

$$f_{\text{circularity}} = \frac{f_{\text{Perimeter}}^2}{f_{\text{Area}}}, \tag{B.45}$$

where f_{Area} and $f_{\text{Perimeter}}$ are the features defined in (B.2) and (B.3) respectively.

B.14 The Normalized Circularity of $\text{Pol}(z)$

The normalized circularity of $\text{Pol}(z)$ is defined as

$$f_{\text{circularity-norm}} = \frac{4 \cdot \pi \cdot f_{\text{Area}}}{f_{\text{Perimeter}}^2}, \tag{B.46}$$

where f_{Area} and $f_{\text{Perimeter}}$ are the features defined in (B.2) and (B.3) respectively.

B.15 The Average Normalized Distance between the Centroid and the Shape Boundary of Pol(z)

The centroid $c = (c_x, c_y)$ of $\mathrm{Pol}(z)$ is defined as

$$c_x = \frac{1}{6 \cdot f_{\mathrm{Area}}} \sum_{i=0}^{M-1} (x_i + x_{[(i+1) \bmod M]})(x_i y_{[(i+1) \bmod M]} - x_{[(i+1) \bmod M]} y_i) \quad (B.47)$$

$$c_y = \frac{1}{6 \cdot f_{\mathrm{Area}}} \sum_{i=0}^{M-1} (y_i + y_{[(i+1) \bmod M]})(x_i y_{[(i+1) \bmod M]} - x_{[(i+1) \bmod M]} y_i) \quad (B.48)$$

where f_{Area} is the feature defined in (B.2). The mean of the normalized distance between the centroid and the shape boundary of $\mathrm{Pol}(z)$ is then calculated as

$$f_{\mathrm{mean-norm-shape}} = \frac{1}{M} \sum_{i=0}^{M-1} \overline{\mathrm{dist}}(v_i, c), \quad (B.49)$$

where

$$\overline{\mathrm{dist}}(v_i, c) = \frac{\mathrm{dist}(v_i, c)}{\mathrm{dist}_{\max}} \quad (B.50)$$

with

$$\mathrm{dist}(v_i, c) = \sqrt{(x_i - c_x)^2 + (y_i - c_y)^2} \quad (B.51)$$

and

$$\mathrm{dist}_{\max} = \max \mathrm{dist}(v_i, c) \ \forall i \quad (B.52)$$

B.16 The Standard Deviation of the Normalized Distances between the Centroid and the Shape Boundary of Pol(z)

The standard deviation of the normalized distances between the centroid and the shape boundary of $P(z)$ is given by

$$f_{\mathrm{std-shape}} = \sqrt{\frac{1}{M} \sum_{i=0}^{M-1} \left(\overline{\mathrm{dist}}(v_i, c) - f_{\mathrm{mean-norm-shape}}\right)^2}, \quad (B.53)$$

where $f_{\mathrm{mean-norm-shape}}$ and $\overline{\mathrm{dist}}(v_i, c)$ are defined in (B.49) and (B.50) respectively.

References

1. Gonzalez, R.C., Wintz, P.A.: Digital Image Processing. Addison-Wesley Publishing Inc., Reading (1987)
2. Haralick, R.M., Shapiro, L.G.: Computer and Robot Vision. Addison-Wesley Publishing Inc., Reading (1992)

3. Loncaric, S.: A survey of shape analysis techniques. Pattern Recognition 31(8), 983–1001 (1998)
4. O'Rourke, J.: Computational Geometry in C, 2nd edn. Cambridge University Press, Cambridge (1998)
5. Russ, J.C.: The Image Processing Handbook. CRC Press, Boca Raton (1992)

Appendix C
Simple Features for People Detection

In this appendix we describe the mathematical definition of the features used for classifying segments into persons. We assume that the robot is equipped with a range sensor that delivers observations consisting of a set of beams in the form

$$z = \{b_0, \ldots, b_{M-1}\} .$$

Each beam b_m corresponds to a tuple (ρ_m, d_m), where ρ_m is the angle of the beam relative to the robot and d_m is the length of the beam.

The beams in the observation scan z are split into subsets using a jump distance condition (cf. Sect. 8.2). We define a segment as a set of N consecutive beams $\mathrm{S} = \{b_0, \ldots, b_{N-1}\}$. Additionally, each beam b_i can be represented by its end-point $p_i = (x_i, y_i)$, with $x_i = d_i \cos \rho_i$ and $y_i = d_i \sin \rho_i$. In this case, we assume that the origin of coordinates lays on the center of the laser sensor.

C.1 Number of Points in the Segment

This feature indicate the number N of points in the segment S.

C.2 The Standard Deviation of the Beam Length

The standard deviation f_σ of the beam length in segment S if defined as

$$f_\sigma = \sqrt{\frac{1}{N} \sum_{i=0}^{N-1} \|p_i - \bar{p}\|^2} , \tag{C.1}$$

where $\bar{p}$ denotes the center of gravity of the segment.

C.3 The Mean Average Deviation from the Median

This feature measures the segment compactness more robustly than the standard deviation. The median of a distribution $f(x)$ is the value where the cumulative distribution function $F(x) = 1/2$. Given an ordered set of K scalar random samples x_i the median $\tilde{x}$ is defined as

$$\tilde{x} = \begin{cases} x_{(K+1)/2} & \text{if } K \text{ is odd} \\ \frac{1}{2}(x_{K/2} + x_{K/2+1}) & \text{if } K \text{ is even}. \end{cases} \tag{C.2}$$

Opposed to the mean, the median is less sensitive to outliers. In our multi-dimensional case, we calculate $\tilde{p}$ using the vector-of-medians approach in [1], i.e. $\tilde{p} = (\tilde{x}, \tilde{y})$. The average deviation from the median is then

$$f_{\varsigma} = \frac{1}{N} \sum_{i=0}^{N-1} \|p_i - \tilde{p}\|. \tag{C.3}$$

C.4 Jump Distance from Preceding Segment

This feature corresponds to the Euclidian distance between the first point of S_i and the last point of S_{i-1}

$$f_{\text{jump-prev}} = \|p_0^{\mathrm{S}_i} - p_{N-1}^{\mathrm{S}_{i-1}}\|. \tag{C.4}$$

C.5 Jump Distance to Succeeding Segment

This feature corresponds to the Euclidian distance between the last point of S_i and the first point of S_{i+1}

$$f_{\text{jump-next}} = \|p_{N-1}^{\mathrm{S}_i} - p_0^{\mathrm{S}_{i+1}}\|. \tag{C.5}$$

C.6 Euclidian Distance between the First and Last Point of a Segment

This feature measures the Euclidian distance between the first and last point of a segment as

$$f_{\text{width}} = \|p_0 - p_{N-1}\|. \tag{C.6}$$

C.7 The Linearity of the Segment

This feature measures the straightness of the segment and corresponds to the residual sum of squares to a line fitted into the segment in the least squares sense. Given the segment points in polar coordinates $p_i = (\rho_i, d_i)$, fitting a line in the Hessian (ϕ, r)-representation that minimizes perpendicular errors from the points onto the line has a closed form solution [2]. Once the line parameters (ϕ, r) are found, the residual sum of squares is calculated as

$$f_{\text{linearity}} = \sum_{i=0}^{N-1} (x_i cos(\phi) + y_i sin(\phi) - r)^2 . \tag{C.7}$$

C.8 The Circularity of the Segment

The circularity of a segment S is calculated by summing up the squared residuals to a fitted circle. Given a set of points in Cartesian coordinates, a fast way to find the best circle in the least squares sense is to parametrize the problem by the vector of unknowns as $x = (x_c \quad y_c \quad x_c^2 + x_c^2 - r_c^2)^T$ where x_c, y_c and r_c denote the circle center and radius. With this, the overdetermined equation system $A \cdot x = b$ can be established as

$$A = \begin{pmatrix} -2x_0 & -2y_0 & 1 \\ -2x_1 & -2y_1 & 1 \\ \vdots & \vdots & \vdots \\ -2x_{N-1} & -2y_{N-1} & 1 \end{pmatrix} \quad b = \begin{pmatrix} -x_0^2 - y_0^2 \\ -x_1^2 - y_1^2 \\ \vdots \\ -x_{N-1}^2 - y_{N-1}^2 \end{pmatrix}, \tag{C.8}$$

and solved using the pseudo-inverse

$$x = (A^T A)^{-1} A^T \cdot b . \tag{C.9}$$

The residual sum of squares is then

$$f_{\text{circularity}} = \sum_{i=0}^{N-1} \left(r_c - \sqrt{(x_c - x_i)^2 + (y_c - y_i)^2} \right)^2 . \tag{C.10}$$

C.9 The Radius of the Circle Fitted to the Segment

This feature represents the radius of the circle fitted to the segment. It corresponds to the value r_c in (C.10).

C.10 The Boundary Length of the Segment

The boundary length f_{boundary} of a segment S measures the length of the poly-line corresponding to the segment and is defined as

$$f_{\text{boundary}} = \sum_{i=0}^{N-2} \| p_i - p_{i+1} \| . \tag{C.11}$$

C.11 The Boundary Regularity of the Segment

The boundary regularity indicates the standard deviation of the distances of adjacent points in a segment and is defined as

$$f_{\text{boundary-reg}} = \sqrt{\frac{1}{N-1}\sum_{i=0}^{N-2}\left(\|(p_i - p_{i+1})\| - \bar{d}\right)^2}\,, \tag{C.12}$$

where $\bar{d}$ is the mean distance between consecutive points.

C.12 The Average Curvature of the Segment

The average curvature $f_{\hat{k}} = \sum \hat{k}_j$ over a segment S is calculated using the following curvature approximation. Given a succession of three points p_A, p_B, and p_C, let A denote the area of the triangle $p_A p_B p_C$ and d_A, d_B, d_C the three distances between the points. Then, an approximation of the discrete curvature of the boundary at p_B is given by

$$f_{\hat{k}} = \frac{4A}{d_A d_B d_C}\,. \tag{C.13}$$

This is an alternative measurement for the radius of the segment, since curvature and radius are inverse proportional.

C.13 The Mean Angular Difference of the Segment

This feature traverses the boundary and calculates the average of the angles β_j between the vectors $\overline{p_{j-1}p_j}$ and $\overline{p_j p_{j+1}}$. The corresponding feature f_{β_j} is then defined as

$$f_{\beta_j} = \angle\left(\overline{p_{j-1}p_j}, \overline{p_j p_{j+1}}\right). \tag{C.14}$$

Care has to be taken that angle differences are properly unwrapped. This features is a measure of the convexity/concavity of segment S.

C.14 Mean Speed between Two Consecutive Scans

Given two scans with their associated timestamps T_k, T_{k+1}, this feature determines the speed v_i for each segment point along its beam as

$$v_i = \frac{d_i^{k+1} - d_i^k}{T_{k+1} - T_k}\,, \tag{C.15}$$

where d_j^k and d_j^{k+1} are the range values of beam j at times k and $k+1$. The final feature f_v averages over all beams in the segment

$$f_v = \sum_{i=0}^{N-1} v_i\,. \tag{C.16}$$

References

1. Aloupis, G.: On Computing Geometric Estimators of Location. PhD thesis, School of Computer Science, McGill University (2001)
2. Arras, K.O.: Feature-Based Robot Navigation in Known and Unknown Environments. PhD thesis, Swiss Federal Institute of Technology Lausanne (EPFL), These No. 2765 (2003)

Springer Tracts in Advanced Robotics

Edited by B. Siciliano, O. Khatib and F. Groen

Further volumes of this series can be found on our homepage: springer.com

Vol. 60: Zhu, W.-H.
Virtual Decomposition Control – Toward Hyper Degrees of Freedom Robots
443 p. 2010 [978-3-642-10723-8]

Vol. 59: Otake, M.
Electroactive Polymer Gel Robots – Modelling and Control of Artificial Muscles
238 p. 2010 [978-3-540-23955-0]

Vol. 58: Kröger, T.
On-Line Trajectory Generation in Robotic Systems – Basic Concepts for Instantaneous Reactions to Unforeseen (Sensor) Events
230 p. 2010 [978-3-642-05174-6]

Vol. 57: Chirikjian, G.S.; Choset, H.; Morales, M., Murphey, T. (Eds.)
Algorithmic Foundations of Robotics VIII – Selected Contributions of the Eighth International Workshop on the Algorithmic Foundations of Robotics
680 p. 2010 [978-3-642-00311-0]

Vol. 56: Buehler, M.; Iagnemma, K.; Singh S. (Eds.)
The DARPA Urban Challenge – Autonomous Vehicles in City Traffic
625 p. 2009 [978-3-642-03990-4]

Vol. 55: Stachniss, C.
Robotic Mapping and Exploration
196 p. 2009 [978-3-642-01096-5]

Vol. 54: Khatib, O.; Kumar, V.; Pappas, G.J. (Eds.)
Experimental Robotics: The Eleventh International Symposium
579 p. 2009 [978-3-642-00195-6]

Vol. 53: Duindam, V.; Stramigioli, S.
Modeling and Control for Efficient Bipedal Walking Robots
211 p. 2009 [978-3-540-89917-4]

Vol. 52: Nüchter, A.
3D Robotic Mapping
201 p. 2009 [978-3-540-89883-2]

Vol. 51: Song, D.
Sharing a Vision
186 p. 2009 [978-3-540-88064-6]

Vol. 50: Alterovitz, R.; Goldberg, K.
Motion Planning in Medicine: Optimization and Simulation Algorithms for Image-Guided Procedures
153 p. 2008 [978-3-540-69257-7]

Vol. 49: Ott, C.
Cartesian Impedance Control of Redundant and Flexible-Joint Robots
190 p. 2008 [978-3-540-69253-9]

Vol. 48: Wolter, D.
Spatial Representation and Reasoning for Robot Mapping
185 p. 2008 [978-3-540-69011-5]

Vol. 47: Akella, S.; Amato, N.; Huang, W.; Mishra, B.; (Eds.)
Algorithmic Foundation of Robotics VII
524 p. 2008 [978-3-540-68404-6]

Vol. 46: Bessière, P.; Laugier, C.; Siegwart R. (Eds.)
Probabilistic Reasoning and Decision Making in Sensory-Motor Systems
375 p. 2008 [978-3-540-79006-8]

Vol. 45: Bicchi, A.; Buss, M.; Ernst, M.O.; Peer A. (Eds.)
The Sense of Touch and Its Rendering
281 p. 2008 [978-3-540-79034-1]

Vol. 44: Bruyninckx, H.; Přeučil, L.; Kulich, M. (Eds.)
European Robotics Symposium 2008
356 p. 2008 [978-3-540-78315-2]

Vol. 43: Lamon, P.
3D-Position Tracking and Control for All-Terrain Robots
105 p. 2008 [978-3-540-78286-5]

Vol. 42: Laugier, C.; Siegwart, R. (Eds.)
Field and Service Robotics
597 p. 2008 [978-3-540-75403-9]

Lightning Source UK Ltd.
Milton Keynes UK
14 March 2010

151358UK00007B/229/P

9 783642 112096